Memoirs of the American Mathematical Society

Number 200

Jonathan Arazy and Yaakov Friedman

Contractive projections in C_1 and C_∞

Published by the

AMERICAN MATHEMATICAL SOCIETY

Providence, Rhode Island

VOLUME 13 · ISSUE 2 · NUMBER 200 (end of volume) · MARCH 1978

ABSTRACT

We characterize all the possible contractive projections and their ranges in the space C_1 of trace class operators on ℓ_2.

There are six mutually different types of elementary subspaces of C_1 on which there are canonical contractive projections from C_1. A subspace X of C_1 is the range of the contractive projection P from C_1 if and only if X is the ℓ_1-sum of elementary subspaces. In this case, there is a unique minimal contractive projection Q from C_1 onto X, and the operator $T = P - Q$ satisfies $\|T\| \leq 1$, $T^2 = 0$ and $T = QT$.

Duality gives a complete description of contractive projections and their ranges in C_∞, the space of all compact operators on ℓ_2.

AMS (MOS) subject classifications (1970). Primary 47D15, 46B99; Secondary 46J10.

Key words and phrases. Trace class operators on Hilbert space, compact operators, contractive projection, support projections, isometry of subspaces, tensor product representation.

Library of Congress Cataloging in Publication Data CIP

Arazy, Jonathan, 1942-
Contractive projections in C_1 and C_∞.

(Memoirs of the American Mathematical Society ; no. 200)
Bibliography: p.
1. Linear operators. 2. Hilbert space.
I. Friedman, Yaakov, joint author. II. Title.
III. Series: American Mathematical Society. Memoirs ; no. 200.
QA3.A57 no. 200 [QA329.2] 510'.8s [515'.7]
ISBN 0-8218-2200-4 77-28610

CONTENTS

The contribution of Jonathan Arazy is a part of his Ph.D. thesis prepared at the Hebrew University of Jerusalem under the supervision of Professor J. Lindenstrauss.

The contribution of Yaakov Friedman is a part of his Ph.D. thesis prepared at Tel-Aviv University under the supervision of Professor A. Lazar.

We wish to thank Professors Lindenstrauss and Lazar for their advice and interest.

CONTRACTIVE PROJECTIONS IN C_1 AND C_∞

CHAPTER 1. INTRODUCTION

Let $B(\ell_2)$ be the Banach space of all bounded operators on ℓ_2 with the usual operator norm $\|x\|_\infty = \sup\{\|x\xi\|_{\ell_2} ; \|\xi\|_{\ell_2} \leq 1\}$, and let C_∞ be its subspace of all compact operators on ℓ_2. C_1, the trace class, is the Banach space of all those operators x on ℓ_2 such that $\|x\|_1 = \text{trace}(x^*x)^{1/2} < \infty$. In this paper we study the contractive (i.e. norm one) projections P in the spaces C_∞ and C_1 and their ranges $R(P)$. Most of our work is done in C_1, and then we get the results for C_∞ by duality. Let us first turn to notations and background material.

The dual of C_∞ is C_1, and the dual of C_1 is $B(\ell_2)$. The duality is given by

$$(1.1) \qquad \langle x,y\rangle = \text{trace}(y^*x)$$

where in the first case $x \in C_\infty$ and $y \in C_1$, and in the second case $x \in C_1$ and $y \in B(\ell_2)$. Thus, the canonical injection of C_∞ in $C_\infty^{**} = B(\ell_2)$ is simply the inclusion operator.

For $x \in B(\ell_2)$ let $x = v(x)\cdot|x|$ be its polar decomposition, where $|x| = (x^*x)^{1/2}$ and $v(x)$ is the unique partial isometry satisfying $x = v(x)|x|$ and $\ker x = \ker v(x)$. The left and right <u>support projections of x</u> are respectively $\ell(x) = v(x)\,v(x)^*$ and $r(x) = v(x)^*v(x)$. These support projections give rise to projections $E(x), F(x)$ and $G(x)$ which are contractive in C_∞, C_1 and $B(\ell_2)$:

$$(1.2) \qquad E(x)y = \ell(x)y\,r(x); \; F(x)y = (1 - \ell(x))\,y(1 - r(x))$$

Received by the editors June 17, 1977.

(1.3) $\quad G(x)y = (1 - \ell(x))y\, r(x) + \ell(x)\, y(1 - r(x))$

Obviously, $1 = E(x) + F(x) + G(x)$. More generally, let X be any subset (in particular, a subspace) of $B(\ell_2)$. Put

(1.4) $\quad \ell(X) = \sup\{\ell(x);\ x \in X\}\ ;\quad r(X) = \sup\{r(x);\ r \in X\}$

and call $\ell(X)$ and $r(X)$ the left and right support projections of X. They also generate projections $E(X)$, $F(X)$ and $G(X)$ as in formulas (1.2) and (1.3).

For each $x \in C_\infty$ the spectral theorem and polar decomposition imply the existence of two orthonormal sequences $\{f_i\}_{i=1}^{\rho(x)}$ and $\{e_i\}_{i=1}^{\rho(x)}$, where $\rho(x) = \dim(\ker x)^\perp$, and a non-increasing sequence of positive numbers $\{s_i\}_{i=1}^{\rho(x)}$ so that the following expansion converges in C_∞ :

(1.5) $$x = \sum_{i=1}^{\rho(x)} s_i(\cdot, f_i)e_i$$

The s_i are the positive eigenvalues of $|x|$ in their multiplicities, and are called the <u>s-numbers</u> of x. Thus, $\|x\|_\infty = s_1$ and $x \in C_1$ if and only if $\|x\|_1 = \sum_{i=1}^{\rho(x)} s_i < \infty$. Note also that $r(x) = \sum_{i=1}^{\rho(x)} (\cdot, f_i)f_i$, $\ell(x) = \sum_{i=1}^{\rho(x)} (\cdot, e_i)e_i$ and $v(x) = \sum_{i=1}^{\rho(x)} (\cdot, f_i)e_i$ where these sums converges in the strong operator topology of $B(\ell_2)$. If we extend $\{f_i\}_{i=1}^{\rho(x)}$ and $\{e_i\}_{i=1}^{\rho(x)}$ to orthonormal bases of ℓ_2, $\{f_i\}_{i\in I}$ and $\{e_i\}_{i\in I}$ respectively, we can represent any $y \in B(\ell_2)$ as matrix $y = (y_{i,j})_{i,j\in I}$ where $y_{i,j} = (yf_j, e_i)$. In this matrix representation x takes the following simple form:

(1.6) $$x = \left(\begin{array}{c|c} \begin{matrix} s_1 & & 0 \\ & \ddots s_i & \\ 0 & & \ddots \end{matrix} & 0 \\ \hline 0 & 0 \end{array}\right)$$

We say that $x,y \in B(\ell_2)$ have <u>disjoint supports</u>, in notation $x \perp y$, if $\ell(x)\,\ell(y) = 0 = r(x)\,r(y)$. This is clearly equivalent to $x^* y = 0 = xy^*$, to $F(x)\,y = y$ and $F(y)\,x = x$ and to the existence of a matrix representation in which

$$(1.7) \qquad x = \begin{pmatrix} \tilde{x} & 0 \\ 0 & 0 \end{pmatrix} ; \quad y = \begin{pmatrix} 0 & 0 \\ 0 & \tilde{y} \end{pmatrix}$$

The notions of disjointness of the supports can be extended to subsets X and Y of $B(\ell_2)$ using the projections $F(X)$, $F(Y)$.

Two elements $x,y \in B(\ell_2)$ are <u>G-related</u> (or, satisfy the <u>G-relation</u>); in notation $x \, G \, y$, if $G(x)y = y$ and $G(y)x = x$. This is clearly equivalent to the existence of a matrix representation in which:

$$(1.8) \qquad x = \begin{pmatrix} 0 & x_1 & 0 \\ x_2 & 0 & 0 \\ 0 & 0 & 0 \end{pmatrix} ; \quad y = \begin{pmatrix} 0 & 0 & y_1 \\ 0 & 0 & 0 \\ y_2 & 0 & 0 \end{pmatrix}$$

where $\ell(x_1) = \ell(y_1)$ and $r(x_2) = r(y_2)$.

For elements of C_1 there is a characterization of the disjointness of the supports of two elements in term of norms:

<u>Proposition 1.1:</u> [7] Let $x,y \in C_1$. Then $x \perp y$ if and only if

$$(1.9) \qquad \|x+y\|_1 + \|x-y\|_1 = 2(\|x\|_1 + \|y\|_1)$$

Another result of this kind is the following variant of [1, Proposition 3.1].

<u>Proposition 1.2:</u> Let $x \in C_1$ be represented as a 2×2 operator matrix:

$$(1.10) \qquad x = \begin{pmatrix} x_{1,1} & x_{1,2} \\ x_{2,1} & x_{2,2} \end{pmatrix} ; \quad x_{i,j} \in C_1$$

Then:

$$(1.11)\qquad \|x\|_1 \geq \Big(\sum_{i,j=1}^{2} \|x_{i,j}\|_1^2\Big)^{1/2}$$

If $x_{i,j} \neq 0$ for $1 \leq i,j \leq 2$ then equality holds in (1.11) if and only if:

$$(1.12)\qquad x_{1,2} = x_{1,1} w^*; \quad x_{2,1} = u\, x_{1,1}; \quad x_{2,2} = u\, x_{1,1} w^*$$

where

$$(1.13)\qquad u = v(x_{2,1})\, v(x_{1,1})^*; \quad w = v(x_{1,2})^*\, v(x_{1,1})$$

If exactly one of the $x_{i,j}$ is zero, or if the only non-zero ones are $x_{1,1}$ and $x_{2,2}$, then equality in (1.11) never holds. If exactly two of the $x_{i,j}$ are not zero, but not $x_{1,1}$ and $x_{2,2}$, then they must relate to each other by formulas analogous to (1.12).

Corollary 1.3: Let $a,b \in B(\ell_2)$ be two projections. Then for $x \in C_1$, $axb = x$ if and only if $\|axb\|_1 = \|x\|_1$.

It follows that for any $x \in C_1$ the projections $E(x)$ and $F(x)$ are strictly contractive in C_1, that is: if $\|E(x)y\|_1 = \|y\|_1$ (respectively, $\|F(x)y\|_1 = \|y\|_1$) then $E(x)y = y$ (respectively, $F(x)y = y$). This conclusion is false for the projection $G(x)$, as the following simple example shows: Let $x = \begin{pmatrix}1 & 0\\ 0 & 0\end{pmatrix}$, $y = \begin{pmatrix}1 & 1\\ 1 & 1\end{pmatrix}$. Then $G(x)y = \begin{pmatrix}0 & 1\\ 1 & 0\end{pmatrix} \neq y$, but $\|y\|_1 = 2 = \|G(x)y\|_1$. There is however a general result which gives in particular the answer to the question "when is $\|G(x)y\|_1 = \|y\|_1$?":

Proposition 1.4: [5, theorem 8.7 in chapter III]

Let $\{a_i\}_{i\in I}$ and $\{b_i\}_{i\in I}$ be two sequences of projections in ℓ_2 so that $a_i a_j = 0 = b_i b_j$ for $i \neq j$. Let $0 \neq x \in C_1$ and put $x_{i,j} = a_i\, x\, b_j$, $i,j \in I$. Let $I_0 \subseteq I$ and put $a = \sum_{i\in I_0} a_i$, $b = \sum_{i\in I_0} b_i$. Then

$$(1.14)\qquad \|x\|_1 \geq \sum_{i\in I_0} \|x_{i,i}\|_1$$

and equality holds in (1.14) if and only if the following conditions hold:

(1.15) $x = a \times b$

(1.16) the partial isometry $v = \sum_{i \in I_0} v(x_{i,i})$ satisfies $v^* x = |x|$.

Note that if (1.15) and (1.16) hold then $x = \sum_{i,j \in I_0} x_{i,j}$, and $r(x_{i,j}) \leq r(x_{j,j})$, $\ell(x_{i,j}) \leq \ell(x_{i,i})$, $x_{i,j} = v(x_{i,i})\, x^*_{j,i}\, v(x_{j,j})$ for every $i,j \in I_0$. It follows for example that if $y = \begin{pmatrix} y_{1,1} & y_{1,2} \\ 0 & y_{2,2} \end{pmatrix}$ and $\|y\|_1 = \|y_{1,1}\|_1 + \|y_{2,2}\|_1$, then $y_{1,2} = 0$.

The space C_1 is not smooth, thus for $0 \neq x \in C_1$ the set

(1.17) $N(x) = \{y \in B(\ell_2);\ \|y\|_\infty \leq 1,\ \langle x,y \rangle = \|x\|_1\}$

contains in general more than one element. As for its structure, we have by the results of [5, Chapter II and III]

Proposition 1.5: Let $0 \neq x \in C_1$, then

(1.18) $N(x) = \{y \in B(\ell_2);\ \ y = v(x) + y_0,\ \|y_0\|_\infty \leq 1,\ y_0 \perp v(x)\}$

We state without proof of the obvious

Proposition 1.6: Let P be a contractive projection in C_1, and let $0 \neq x \in R(P)$. Then $P^*(N(x)) \subseteq N(x)$.

Another result which is of importance in studying isometrical problems in C_1 is the following

Proposition 1.7: [2] Let Z be a Banach space so that Z^* is separable, and let K be a norm-closed, convex, bounded subset of Z^*. Then K is the norm-closed convex hull of its extreme point.

This proposition can be applied to subsets K of $C_1 = C^*_\infty$. In particular if X is a closed subspace of C_1 then its unit ball is the closed convex hull of the set of its extreme points, Ext B_X. If $\{0\} \neq X$ we get that Ext $B_X \neq \phi$.

Let us mention some results on tensor products of operators (for the basic properties of this notion, see [8]). Here we use the notations $C_\infty(H)$ and $C_1(H)$ for the spaces of compact and trace-class operators on the Hilbert space H.

Let $\ell_2 \otimes \ell_2$ be the Hilbert-space tensor product of ℓ_2 with itself (in which the inner product is defined by $(\xi_1 \otimes \xi_2, \eta_1 \otimes \eta_2) = $ $= (\xi_1,\eta_1)\cdot(\xi_2,\eta_2)$ and by linearity). If $x,y \in B(\ell_2)$ then there exists a unique $x \otimes y \in B(\ell_2 \otimes \ell_2)$ satisfying $(x \otimes y)(\xi \otimes \eta) = x\xi \otimes y\xi$ for every $\xi,\eta \in \ell_2$, and

$$(1.19) \qquad \|x \otimes y\|_\infty = \|x\|_\infty \cdot \|y\|_\infty$$

If $x,y \in C_\infty(\ell_2)$, then $x \otimes y \in C_\infty(\ell_2 \otimes \ell_2)$. If $x,y \in C_1(\ell_2)$, then $x \otimes y \in C_1(\ell_2 \otimes \ell_2)$ and

$$(1.20) \qquad \|x \otimes y\|_1 = \|x\|_1 \cdot \|y\|_1$$

Moreover, $C_p(\ell_2 \otimes \ell_2) = \overline{\text{span}}\,\{x \otimes y;\ x,y \in C_p(\ell_2)\}$ $(p = 1,\infty)$, and $B(\ell_2 \otimes \ell_2) = \overline{\text{span}}\,\{x \otimes y;\ x,y \in B(\ell_2)\}$, where in the last formula the closure is taken in the weak operator topology. We denote $C_p(\ell_2 \otimes \ell_2)$ by $C_p \otimes C_p$ and $B(\ell_2 \otimes \ell_2)$ by $B(\ell_2) \otimes B(\ell_2)$. Let $\{e_i\}_{i=1}^\infty$ and $\{f_i\}_{i=1}^\infty$ be two orthonormal bases of ℓ_2, let $N = \bigcup_{k=1}^\infty A_k = \bigcup_{k=1}^\infty B_k$ be two partitions of the set N of positive integers into pairwise disjoint infinite subsets, and let $\phi_k : N \to A_k$, $\psi_k : N \to B_k$ be one-to-one and onto mappings, $1 \leq k < \infty$. Since $\{f_\ell \otimes f_j\}_{\ell,j=1}^\infty$ and $\{e_k \otimes e_i\}_{k,i=1}^\infty$ are orthonormal bases for $\ell_2 \otimes \ell_2$, there exist isometries u,w of ℓ_2 onto $\ell_2 \otimes \ell_2$ so that for every $i,j,k,\ell \in N$:

(1.21) $w\, f_{\psi_\ell(j)} = f_\ell \otimes f_j$; $u\, e_{\phi_k(i)} = e_k \otimes e_i$

Define for $x \in B(\ell_2)$

(1.22) $Vx = u\, x\, w^{-1}$

Then V is an isometry of $B(\ell_2)$ onto $B(\ell_2) \otimes B(\ell_2)$, and an isometry of C_p onto $C_p \otimes C_p$ $(p = 1,\infty)$.

In the sequel we shall therefore identify $B(\ell_2)$ with $B(\ell_2) \otimes B(\ell_2)$ and C_p with $C_p \otimes C_p$ $(p = 1,\infty)$; the identification will always be made using some given orthonormal bases $\{e_i\}_{i=1}^{\infty}$ and $\{f_i\}_{i=1}^{\infty}$ of ℓ_2, some partitions $N = \bigcup_{k=1}^{\infty} A_k = \bigcup_{k=1}^{\infty} B_k$ of the positive integers and some mappings ϕ_k and ψ_k as above by means of formulas (1.21) and (1.22). We call this identification a "tensor-product representation" of C_p and $B(\ell_2)$. Usually it will be clear from the context (or immaterial) how the identification is made. For example, let $x \in C_1$ and $y,z \in B(\ell_2)$. Formally, the expression $\langle x, y \otimes z\rangle$ is meaningless since x acts on ℓ_2 and $y \otimes z$ acts on $\ell_2 \otimes \ell_2$. However, if we choose $\{e_i\}_{i=1}^{\infty}$, $\{f_i\}_{i=1}^{\infty}$, $\{A_k\}_{k=1}^{\infty}$, $\{B_k\}_{k=1}^{\infty}$, $\{\phi_k\}_{k=1}^{\infty}$ and $\{\psi_k\}_{k=1}^{\infty}$ and construct V by formulas (1.21) and (1.22) then $\langle Vx, y \otimes z\rangle$ is defined (and, of course, depends on V). Thus, if V is understood, then $\langle x, y \otimes z\rangle$ will simply stand for $\langle Vx, y \otimes z\rangle$.

The use of tensor products of operators enables us to denote in a convenient manner large operator-matrices, and formulas involving many partial isometries. To explain this, let $\ell_2 = \sum_{i=1}^{\infty} \oplus K_i = \sum_{i=1}^{\infty} \oplus H_i$ be two representations of ℓ_2 as a direct sum of pairwise orthogonal infinite-dimensional closed subspaces. Let $\{e_i\}_{i\in A_k}$, $\{f_j\}_{j\in B_k}$ be orthonormal bases of K_k and H_k respectively, where for $\ell \neq k$ $A_k \cap A_\ell = \phi = B_k \cap B_\ell$, and $\bigcup_{k=1}^{\infty} A_k = \bigcup_{k=1}^{\infty} B_k = N$. Choose one-to-one and onto mappings $\phi_k : N \to A_k$ and $\psi_k : N \to B_k$, $1 \leq k < \infty$, and let u,w and V

be defined by means of (1.21) and (1.22). Let $x \in B(\ell_2)$ and let $x_{k,\ell}$ be the operator from H_ℓ into K_k induced by x. The operator-matrix $(x_{k,\ell})_{k,\ell=1}^{\infty}$ describes of course the action of x on ℓ_2 in the usual sense. Now $Vx = \sum_{k,\ell=1}^{\infty} e_{k,\ell} \otimes x_{k,\ell}$, where $e_{k,\ell}$ is the one rank operator defined by $e_{k,\ell} = (\cdot, f_\ell)e_k$. Thus, if V is understood then the expression $x = \sum_{k,\ell=1}^{\infty} e_{k,\ell} \otimes x_{k,\ell}$ means simply that x has a matrix with respect to the given decompositions of ℓ_2, whose component in the (k,ℓ)-place is the operator $x_{k,\ell}$. A typical use is the following. Let $\{K_i\}_{i=1}^{\infty}$ and $\{H_i\}_{i=1}^{\infty}$ be as before, and let $\{u_i\}_{i=1}^{\infty}$ and $\{w_i\}_{i=1}^{\infty}$ be partial isometries in ℓ_2 so that u_i takes K_1 isometrically onto K_i and vanishes on $K_1^{\perp}$, and w_i takes H_1 isometrically onto H_i and vanishes on $H_1^{\perp}$, $1 \leq i < \infty$. If x_0 is some operator from H_1 to K_1, then $x_{k,\ell} = u_k x_0 w_\ell^*$ is the operator from H_ℓ to K_k, whose action is the same as that of x_0, modulo the partial isometries u_k and w_ℓ. Using the above constructed V, we have $Vx_{k,\ell} = e_{k,\ell} \otimes x_0$. For convenience, however, we write in such a situation also $x_{k,\ell} = e_{k,\ell} \otimes x_0$. Thus, for example, the conclusion of proposition 1.2 in case of equality can be written simply as $x_{i,j} = e_{ij} \otimes x_o$, $1 \leq i, j \leq 2$.

Our program is now the following. In section 2 we describe what we call "elementary subspaces" of C_p $(p = 1,\infty)$ and prove that on each such subspace there is a canonical contractive projection from C_p. In the end of this section we formulate the main result of this work in three theorems: 2.14, 2.15 and 2.16. The description of the three first types of elementary subspaces is quite simple, but the description of the elementary subspaces of types 4, 5 and 6 is rather complicated. Theorems 2.14, 2.15 and 2.16 and most of our results in sections 3 and 4 can however be fully understood without any reference to the description of elementary subspaces of types 4, 5 and 6. In sections 3, 4 and 5 we work in C_1. Section 3 contains some results on the relations between P and the projections $E(x)$, $F(x)$ and $G(x)$ for $x \in R(P)$. In this section we introduce also the notion of an atom of P

and study some of its properties. Section 4 is devoted to the study of the possible relations between atoms of P, and in section 5 we show how the range of P is constructed using its atoms. In the end of section 5 we combine the results of sections 3, 4 and 5 and prove our main theorems for C_1; and then by duality also for C_∞. In section 6 we study the isometries of elementary subspaces of C_1 into C_1, and we show that the six types of elementary subspaces of C_p $(p = 1,\infty)$ introduced in section 2 are indeed different. Finally, section 7 contains some corollaries, concluding remarks and open problems. In particular we prove in this section that there is no monotone basis in either C_∞ or C_1.

Several of our results, are trivially true also in the spaces C_p of all operators x on ℓ_2 with $\|x\|_p = (\mathrm{trace}(x^*x)^{p/2})^{1/p} < \infty$, and even in any symmetric norm ideal (see [5] for the definition). If $p = 2$, then C_2 is well known to be a Hilbert space (its elements are called "Hilbert Schmidt operators") so every closed subspace is the range of a unique orthogonal (and thus - contractive) projection from C_2. For $1 < p \neq 2 < \infty$ we know the full description of contractive projection and their ranges only in $C_p(\ell_2^n), n < \infty$. The situation is analogous to that in C_1, with only minor changes (for example, here a subspace is the range of at most <u>one</u> contractive projection from C_p, while in C_1 there may be several contractive projections on a given subspace). Unfortunately, our techniques in proving our results for $C_p(\ell_2^n)$ lead to some difficulties in the infinite dimensional case that we cannot overcome as yet.

There is a similarity between known facts concerning contractive projections and their ranges in L_1-spaces (see [3]), and the results we obtained in this work for C_1. Some ideas in the proofs are the same in L_1 and C_1, but the proofs themselves are different, and the case of C_1 is more involved because of the non-commutativity of the multiplication.

In this work we use complex scalars. The difference between the real and the complex case will be explained in section 7. Also, by a "projection in ℓ_2" we will always mean a "selfadjoint (i.e. orthogonal) projection"; no other projections in ℓ_2 will be considered.

2. DESCRIPTION OF ELEMENTARY SUBSPACES AND FORMULATION OF THE MAIN RESULTS.

a; Types 1, 2 and 3: Elementary subspaces of C_p of the forms $C_p^{n,m}(y,z)$, $SY_p^m(x)$ and $A_p^m(x)$ $(p = 1,\infty)$

Let $\{e_i\}_{i=1}^\infty$ and $\{f_j\}_{j=1}^\infty$ be two orthonormal bases of ℓ_2, and let $e_{i,j} = (\cdot,f_j)e_i$, $1 \leq i,j < \infty$. In the matrix representation associated with these bases the matrix of $e_{i,j}$ has "1" in the (i,j)-place and zeros elsewhere. We therefore call $\{e_{i,j}\}_{i,j=1}^\infty$ the "standard unit matrices" associated with the pair of bases $(\{e_i\}_{i=1}^\infty, \{f_j\}_{j=1}^\infty)$, and denote $C_p^{n,m} = \overline{\text{span}}\,\{e_{i,j}\}_{\substack{1\leq j\leq n\\ 1\leq j\leq m}}$, $1 \leq n,m \leq \infty$, the closure is taken in C_p $(p = 1,\infty)$. Usually, the bases $\{e_i\}_{i=1}^\infty$ and $\{f_j\}_{j=1}^\infty$ are understood and the reference to them is omitted. We also put

$$(2.1) \qquad s_{i,j} = \begin{cases} e_{i.i} \ ; & 1 \leq i = j < \infty \\ e_{i,j} + e_{j,i} \ ; & 1 \leq i < j < \infty \end{cases}$$

and

$$(2.2) \qquad a_{i,j} = e_{i,j} - e_{j,i} \ ; \quad 1 \leq i < j < \infty \ ;$$

and call the $\{s_{i,j}\}_{1\leq i\leq j<\infty}$ the "elementary symmetric matrices", and the $\{a_{i,j}\}_{1\leq i<j<\infty}$ the "elementary antisymmetric matrices" associated with the pair of bases $(\{e_i\}_{i=1}^\infty, \{f_i\}_{i=1}^\infty)$. For $p = 1, \infty$ and $m \leq \infty$ we denote $SY_p^m = \overline{\text{span}}\,\{s_{i,j}\ ;\ 1 \leq i \leq j \leq m\}$ and $A_p^m = \overline{\text{span}}\,\{a_{i,j};\ 1 \leq i < j \leq m\}$ the subspaces of $C_p^{m,m}$ consisting of all symmetric and anti-symmetric matrices respectively. The spaces $C_p^{\infty,\infty}$, SY_p^∞ and A_p^∞ are denoted also by C_p, SY_p and A_p respectively.

We define now our first three types of elementary subspaces of

C_p $(p = 1,\infty)$. This definition depends on our fixed but arbitrary bases $\{e_i\}_{i=1}^{\infty}$ and $\{f_i\}_{i=1}^{\infty}$ of ℓ_2. Also, we identify ℓ_2 with $\ell_2 \otimes \ell_2$ and C_p with $C_p \otimes C_p$ $(p = 1,\infty)$ (and $B(\ell_2)$ with $(B(\ell_2)\otimes B(\ell_2))$, using these bases in the standard way described in the introduction. Such a comment will be frequently omitted in the sequel.

Definition 2.1: Let $p = 1$ or $p = \infty$. If $y,z \in C_p$, $y \perp z$ and $\|y+z\|_p = 1$ and if $1 \leq n,m \leq \infty$ we define

$$(2.3) \qquad C_p^{n,m}(y,z) = \overline{\text{span}}\,\{e_{i,j} \otimes y + e_{j,i} \otimes z\}_{\substack{1\leq i\leq n \\ 1\leq j\leq m}} .$$

If $2 \leq n \leq m \leq \infty$ and $3 \leq m$, <u>$C_p^{n,m}(y,z)$ is called an elementary subspace of C_p of type 1.</u>

If $x \in C_p$, $\|x\|_p = 1$, then for every $1 \leq m \leq \infty$ we put

$$(2.4) \qquad SY_p^m(x) = \overline{\text{span}}\,\{s_{i,j} \otimes x\}_{1\leq i\leq j\leq m} = SY_p^m \otimes x$$

and call <u>$SY_p^m(x)$ an elementary subspace of C_p of type 2.</u>

If $2 \leq m \leq \infty$ and $x \in C_p$ with $\|x\|_p = 1$, we put

$$(2.5) \qquad A_p^m(x) = \overline{\text{span}}\,\{a_{i,j} \otimes x\}_{1\leq i<j\leq m} = A_p^m \otimes x,$$

and for $5 \leq m \leq \infty$ we call <u>$A_p^m(x)$ an elementary subspace of C_p of type 3.</u>

The spaces $C_p^{\infty,\infty}(y,z)$, $SY_p^{\infty}(x)$ and $A_p^{\infty}(x)$ are denoted also by $C_p(y,z)$, $SY_p(x)$ and $A_p(x)$ respectively.

In the definition above our restrictions on n,m were posed because we want the classes of elementary subspaces to be disjoint. Thus the spaces $C_p^{n,m}(y,z)$ for $\min\{n,m\} = 1$ or for $n = m = 2$, and the spaces $A_p^m(x)$ for $2 \leq m \leq 4$, are not considered as elementary subspaces of C_p of types 1 and 3 respectively. The spaces $C_p^{1,1}(y,z)$ and $A_p^2(x)$ are one-dimensional and thus "coincide" with $SY_p^1(x)$. The spaces $C_p^{n,1}(y,z)$, $C_p^{1,m}(y,z)$ and $A_p^3(x)$ are elementary Hilbert subspaces of C_p and will be classified among the elementary subspaces of C_p of type 4. Finally, $C_p^{2,2}(y,z)$ and $A_p^4(x)$ will be classified among the elementary subspaces of C_p of type 5.

Proposition 2.2: Let $C_p^{n,m}(y,z)$, $SY_p^m(x)$ and $A_p^m(x)$ be as in definition 2.1. Then they are isometric in the natural way to $C_p^{n,m}$, SY_p^m and A_p^m respectively, and there is a contractive projection from C_p onto each of them.

Proof: The mappings $e_{i,j} \to e_{i,j} \otimes y + e_{j,i} \otimes z$, $s_{i,j} \to s_{i,j} \otimes x$ and $a_{i,j} \to a_{i,j} \otimes x$ extend to linear isometries of $C_p^{n,m}$, SY_p^m and A_p^m onto $C_p^{n,m}(y,z)$, $SY_p^m(x)$ and $A_p^m(x)$ respectively. This is due to the formula $\|a \otimes b\|_p = \|a\|_p \|b\|_p$ $(p = 1,\infty)$, and the fact that if $a_1 \perp a_2$ and b_1,b_2 are arbitrary, then $(a_1 \otimes b_1) \perp (a_2 \otimes b_2)$. If $p = \infty$ the above isometries acts isometrically also in the corresponding subspaces of $B(\ell_2)$ (in this case $B^{n,m}(\ell_2)(y,z)$ will denote the closure in the weak operator topology of span $\{e_{i,j} \otimes y + e_{j,i} \otimes z\}_{\substack{1\le i\le n\\ 1\le j\le m}}$ and so on).

In [1] it was porved that there are canonical contractive projections of C_p onto $C_p^{n,m}(y,z)$, $p = 1,\infty$. Let us just write down the formulas of these "canonical" projections for future reference. For $p = 1$ and $a \in C_1$ we define:

$$(2.6) \qquad Pa = \sum_{\substack{1\le i\le n\\ 1\le j\le m}} \langle a, e_{i,j} \otimes v(y) + e_{j,i} \otimes v(z) \rangle (e_{i,j} \otimes y + e_{j,i} \otimes z)$$

If $p = \infty$ and for example $\|y\|_\infty = 1$ we choose some $y_0 \in C_1$ with $\|y_0\|_1 = 1 = \langle y,y_0\rangle$ (and thus necessarily $y_0 \perp z$), and define for $a \in C_\infty$

$$(2.7) \qquad Pa = \sum_{\substack{1\le i\le n\\ 1\le j\le m}} \langle a, e_{i,j} \otimes y_0 \rangle (e_{i,j} \otimes y + e_{j,i} \otimes z)$$

In the proof of the contractiveness of the projections (2.6) and (2.7) we make use of the following fact which will be used again below: if $a,b \in C_1$ then trace $(a \otimes b) = \text{trace}(a) \cdot \text{trace}(b)$, and thus for every $c,d \in B(\ell_2)$; $\langle c \otimes d, a \otimes b\rangle = \langle c,a\rangle \cdot \langle d,b\rangle$.

For $p = 1$ the "canonical" contractive projections of C_1 onto $SY_1^m(x)$ and $A_1^m(x)$ are respectively:

$$\text{(2.8)} \quad Qa = \sum_{1 \le i \le j \le m} \langle a, s_{i,j} \otimes v(x)\rangle \, (s_{i,j} \otimes x)/\|s_{i,j}\|_1$$

and

$$\text{(2.9)} \quad Ra = \sum_{1 \le i < j \le m} \langle a, a_{i,j} \otimes v(x)\rangle \, (a_{i,j} \otimes x)/2$$

We show for example that Q is contractive (the proof for R is almost the same):

$$\text{(2.10)} \quad \|Qa\|_1 = \Big\| \sum_{1 \le i \le j \le m} \langle a, s_{i,j} \otimes v(x)\rangle\, s_{i,j}/\|s_{i,j}\|_1 \Big\|_1$$

$$= \sup_{\substack{b \in B(\ell_2) \\ \|b\|_\infty = 1}} \Big| \Big\langle \sum_{1 \le i \le j \le m} \langle a, s_{i,j} \otimes v(x)\rangle\, s_{i,j}/\|s_{i,j}\|_1 , b \Big\rangle \Big|$$

$$= \sup_{\substack{b \in B(\ell_2) \\ \|b\|_\infty = 1}} \Big| \Big\langle a, \sum_{1 \le i \le j \le m} \Big\langle b, \frac{s_{i,j}}{\|s_{i,j}\|_1} \Big\rangle\, s_{i,j} \otimes v(x) \Big\rangle \Big|$$

$$\le \|a\|_1 \sup_{\substack{b \in B(\ell_2) \\ \|b\|_\infty = 1}} \Big\| \sum_{1 \le i \le j \le m} \Big\langle b, \frac{s_{i,j}}{\|s_{i,j}\|_1} \Big\rangle\, s_{i,j} \Big\|_\infty$$

$$\le \|a\|_1 \sup_{\substack{b \in B(\ell_2) \\ \|b\|_\infty = 1}} \|(b+b^T)/2\|_\infty = \|a\|_1$$

(in the last step we use the trivial fact $\|b\|_\infty = \|b^T\|_\infty$ where b^T is the transpose of b).

If $p = \infty$ we choose $x_0 \in C_1$ with $\|x_0\|_1 = 1 = \langle x, x_0 \rangle$ and define for $a \in C_\infty$:

$$Qa = \sum_{1 \leq i \leq j \leq m} \langle a, s_{i,j} \otimes x_0 \rangle (s_{i,j} \otimes x)/\|s_{i,j}\|_1$$

and

$$(2.12) \qquad Ra = \sum_{1 \leq i < j \leq m} \langle a, a_{i,j} \otimes x_0 \rangle (a_{i,j} \otimes x)/2 .$$

Then Q and R are contractive projections of C_∞ onto $SY^m_\infty(x)$ and $A^m_\infty(x)$ respectively, the proof is very similar to (2.10) and thus we omit it. We call such Q (respectively R) a "canonical" projection of C_∞ onto $SY^m_\infty(x)$ (resp. $A^m_\infty(x)$). □

Notations

The description of the next three elementary types of ranges of contractive projections in C_p $(p = 1,\infty)$ is more complicated, and we use in it the following notations: $I = \{1,2,3,\ldots\}$ is the set of all positive integers, $\mathscr{P}$ is the set of all finite subsets of I (including ϕ). By α,β,γ we denote elements of $\mathscr{P}$, and $I_n = \{1,2,\ldots,n\}$. Every matrix x will be infinite and its coordinates are indexed by $\mathscr{P}$, that is: $x = (x_{\alpha,\beta})_{\alpha,\beta\in\mathscr{P}}$. We identify matrices $x = (x_{\alpha,\beta})_{\alpha,\beta\subseteq I_n}$ with matrices indexed by $\mathscr{P}$ whose (α,β)-entry vanish if $\alpha \nsubseteq I_n$ or $\beta \nsubseteq I_n$. As usual, $|\alpha|$ will denote the cardinality of $\alpha \in \mathscr{P}$. We continue to use orthonormal bases $\{e_\alpha\}_{\alpha\in\mathscr{P}}, \{f_\alpha\}_{\alpha\in\mathscr{P}}$ of ℓ_2, and denote by $e_{\alpha,\beta} = (\cdot, f_\beta)e_\alpha$ the standard unit matrices associated with the pair of bases $(\{e_\alpha\}_{\alpha\in\mathscr{P}}, \{f_\alpha\}_{\alpha\in\mathscr{P}})$. If $\alpha_0,\beta_0 \in \mathscr{P}$ and $x = (x_{\alpha,\beta})_{\alpha,\beta\in\mathscr{P}}$, we define a matrix $x^{(\alpha_0,\beta_0)}$ by the formula:

$$(2.13)\qquad (x^{(\alpha_0,\beta_0)})_{\alpha,\beta} = \begin{cases} x_{\alpha \sim \alpha_0, \beta \sim \beta_0} ; & \text{if } \beta_0 \subseteq \beta \text{ and } \alpha_0 \subseteq \alpha \\ 0 ; & \text{otherwise} \end{cases}$$

Also, $x^{(\alpha_0)}$ stands for $x^{(\alpha_0,\alpha_0)}$ and $x^{(n)}$ for $x^{(\{n\})}$. Note that if $\alpha_0,\beta_0,\gamma_0 \in \mathscr{P}$ and x,y are arbitrary matrices, then

$$(2.14)\qquad x^{(\alpha_0,\beta_0)}\, y^{(\beta_0,\gamma_0)} = (xy)^{(\alpha_0,\gamma_0)}$$

The proof is straightforward. Also if $y = (y_{\alpha,\beta})_{\alpha,\beta\subseteq I_{n-1}}$ and x is an arbitrary matrix, then $y \perp x^{(n)}$.

Finally, if $k \in \alpha \subseteq I$, we put $i(\alpha,k) = |\{k' \in \alpha; k' < k\}|$. Thus, if $\alpha = \{\alpha_1,\alpha_2,\ldots,\alpha_n\}$ with $\alpha_i < \alpha_{i+1}$, and if $i(\alpha,k) = i_0$, then $k = \alpha_{i_0+1}$.

<u>b. Type 4:</u> <u>Elementary Hilbert subspaces of C_p $(p = 1,\infty)$</u>

For $n \in I$ and $k,m \in I_n$ we define

$$(2.15)\qquad x_{n,k,m} = \sum_{\substack{k\in\alpha\subseteq I_n \\ |\alpha|=m}} (-1)^{i(\alpha,k)}\, e_{\alpha,\alpha\sim\{k\}}$$

If $n = \infty$ we define the $x_{n,k,m}$ only for $m = 1$ and $m = \infty$, where $k \in I$:

$$(2.16)\qquad x_{\infty,k,\infty} = e_{\phi,\{k\}} \;;\quad x_{\infty,k,1} = e_{\{k\},\phi}$$

We give now a few examples of the matrices $A_{n,m} = \sum_{k=1}^{n} a_k\, x_{n,k,m}$. The indices range over $\mathscr{P}_n$, the set of all subsets of I_n, in the following ordering: for every $\phi\neq\alpha \in \mathscr{P}_n$ we define $\phi < \alpha$. Let us continue inductively.

(2.17) $\left\{\begin{array}{l}\text{Let } \alpha = \{\alpha_1,\alpha_2,\ldots,\alpha_k\} \text{ and } \beta = \{\beta_1,\beta_2,\ldots,\beta_m\} \text{ be two} \\ \text{different members of } \mathscr{P}_n \text{ where } \alpha_i < \alpha_{i+1} \text{ and } \beta_j < \beta_{j+1} \\ \text{for every } i \text{ and } j. \text{ If } \{\alpha_1,\ldots,\alpha_{k-1}\} < \{\beta_1,\ldots,\beta_{m-1}\} \text{ we} \\ \text{put } \alpha < \beta. \text{ If } k = m \text{ and } \{\alpha_1,\ldots,\alpha_{k-1}\} = \{\beta_1,\ldots,\beta_{m-1}\} \\ \text{and } \alpha_k < \beta_m, \text{ we put } \alpha < \beta.\end{array}\right.$

Example 2.3

$n = 2$

(2.18) $a_1x_{2,1,1}+a_2x_{2,2,1}+b_1x_{2,1,2}+b_2x_{2,2,2} =$

a_1			
a_2			
	$-b_2$	b_1	

$n = 3$

(2.19) $\sum_{k=1}^{3} a_k\, x_{3,k,1} +$

$+ \sum_{k=1}^{3} b_k\, x_{3,k,2} + \sum_{k=1}^{3} c_k\, x_{3,k,3} =$

a_1							
a_2							
a_3							
	$-b_2$	b_1					
	$-b_3$		b_1				
		$-b_3$	b_2				
				c_3	$-c_2$	c_1	

n = 4

$$(2.20)\quad \sum_{k=1}^{4} a_k x_{4,k,1} + \sum_{k=1}^{4} b_k\, x_{4,k,2} + \sum_{k=1}^{4} c_k\, x_{4,k,3} + \sum_{k=1}^{4} d_k\, x_{4,k,4} =$$

a_1															
a_2															
a_3															
a_4															
	$-b_2$	b_1													
	$-b_3$		b_1												
	$-b_4$			b_1											
		$-b_3$	b_2												
		$-b_4$		b_2											
			$-b_4$	b_3											
					c_3	$-c_2$		c_1							
					c_4		$-c_2$		c_1						
						c_4	$-c_3$			c_1					
								c_4	$-c_3$	c_2					
											$-d_4$	d_3	$-d_2$	d_1	

Note that for $m \neq m'$ and any k,k' we have $x_{n,k,m} \perp x_{n,k',m'}$.

Proposition 2.4: Fix $1 \leq m \leq n \leq \infty$ and $p = 1,\infty$, and let $h_k = x_{n,k,m}/\|x_{n,k,m}\|_p$. Then for every scalar $\{a_k\}_{k=1}^n$ we have

$$(2.19)\quad \Big\|\sum_{k=1}^{n} a_k h_k\Big\|_p = \Big(\sum_{k=1}^{n} |a_k|^2\Big)^{1/2}$$

Consequently, the "canonical" projection from C_p onto $[h_k]_{k=1}^n$ is contractive.

Proof: For convenience, we assume without loss of generality that our two orthonormal bases for ℓ_2 are the same, i.e. $e_\alpha = f_\alpha$ for every $\alpha \in \mathscr{P}$.

We treat first the simplest cases $m = n$ and $m = 1$ where $1 \leq n \leq \infty$. Since the latter can be obtained from the first by transposing the matrices and using suitable unitary multiplications, we treat only the first case. Let $\{a_k\}_{k=1}^n$ be scalars, then

$$(2.22) \qquad (\sum_k a_k h_k) \cdot (\sum_k a_k h_k)^* = \sum_k |a_k|^2 \cdot e_{\phi,\phi}$$

So $\sum_k a_k h_k$ is one-rank operator with one s-number $(\sum_k |a_k|^2)^{1/2}$. Thus:

$$(2.23) \qquad \|\sum_k a_k h_k\|_p = \|((\sum_k a_k h_k)(\sum_k a_k h_k)^*)^{1/2}\|_p = (\sum_k |a_k|^2)^{1/2} .$$

Let

$$(2.24) \qquad Px = \sum_k \langle x,h_k\rangle h_k .$$

Then:

$$(2.25) \qquad \|Px\|_p^2 = \sum_k |\langle x,h_k\rangle|^2 = \langle x,Px\rangle \leq \|x\|_p \|Px\|_{p^*} = \|x\|_p \|Px\|_p$$

(where p^* is the conjugate index of p).

Assume now that $1 < m < n < \infty$, let $\{a_k\}_{k=1}^n$ be scalars with $\lambda = \sum_{k=1}^n |a_k|^2 > 0$. Put $A_{n,m} = \sum_{k=1}^n a_k x_{n,k,m}$

Claim:

$$(2.26) \qquad A_{n,m}^* A_{n,m} + A_{n,m-1} A_{n,m-1}^* = \lambda \sum_{\substack{\alpha \subseteq I_n \\ |\alpha|=m-1}} e_{\alpha,\alpha}$$

Indeed, by an obvious computation:

$$(2.27) \quad A_{n,m}^* A_{n,m} = \sum_{k,k'=1}^n \bar{a}_k a_{k'} \sum_{\substack{\alpha \subseteq I_n \\ |\alpha|=m \\ k,k' \in \alpha}} (-1)^{i(\alpha,k)+i(\alpha,k')} e_{\alpha \sim \{k\},\, \alpha \sim \{k'\}}$$

$$= \sum_{k=1}^{n} |a_k|^2 \sum_{\substack{\alpha \subseteq I_n \\ |\alpha|=m \\ k\in\alpha}} e_{\alpha \sim \{k\},\, \alpha \sim \{k\}}$$

$$+ \sum_{\substack{k,k'=1 \\ k\neq k'}}^{n} \bar{a}_k a_{k'} \sum_{\substack{\alpha \subseteq I_n \\ |\alpha|=m \\ k,k'\in\alpha}} (-1)^{i(\alpha,k)+i(\alpha,k')} e_{\alpha \sim \{k\},\, \alpha \sim \{k'\}}$$

And also

$$(2.28) \quad A_{n,m-1} A^*_{n,m-1} =$$

$$= \sum_{k,k'=1}^{n} \bar{a}_k a_{k'} \sum_{\substack{\alpha \subseteq I_n \\ |\alpha|=m-1 \\ k'\in\alpha}} \sum_{\substack{\beta \subseteq I_n \\ |\beta|=m-1 \\ k\in\beta}} (-1)^{i(\alpha,k')+i(\beta,k)} \cdot e_{\alpha,\alpha \sim \{k'\}} e_{\beta\sim\{k\},\beta}$$

$$= \sum_{k=1}^{n} |a_k|^2 \sum_{\substack{\alpha \subseteq I_n \\ |\alpha|=m-1 \\ k\in\alpha}} e_{\alpha,\alpha}$$

$$+ \sum_{\substack{k,k'=1 \\ k\neq k'}}^{n} \bar{a}_k a_{k'} \sum_{\substack{\gamma \subseteq I_n \sim\{k,k'\} \\ |\gamma|=m-2}} (-1)^{i(\gamma\cup\{k'\},k')+i(\gamma\cup\{k\},k)} e_{\gamma\cup\{k'\},\gamma\cup\{k\}}$$

If $k,k' \in \alpha \subseteq I_n$, $k \neq k'$ and $|\alpha| = m$, and if $\gamma = \alpha \sim \{k,k'\}$, then

$$(2.29)\qquad i(\alpha,k) + i(\alpha,k') = i(\gamma\cup\{k'\},k') + i(\gamma\cup\{k\},k) + 1.$$

So

$$(2.30)\qquad A^*_{n,m} A_{n,m} + A_{n,m-1} A^*_{n,m-1} =$$

$$= \sum_{k=1}^n |a_k|^2 \sum_{\substack{\alpha\subseteq I_n\\ |\alpha|=m\\ k\in\alpha}} e_{\alpha\sim\{k\},\,\alpha\sim\{k\}} + \sum_{k=1}^n |a_k|^2 \sum_{\substack{\alpha\subseteq I_n\\ |\alpha|=m-1\\ k\in\alpha}} e_{\alpha,\alpha}$$

$$= \sum_{k=1}^n |a_k|^2 \Big(\sum_{\substack{\alpha\subseteq I_n\\ |\alpha|=m-1\\ k\notin\alpha}} e_{\alpha,\alpha} + \sum_{\substack{\alpha\subseteq I_n\\ |\alpha|=m-1\\ k\in\alpha}} e_{\alpha,\alpha}\Big) = \lambda \sum_{\substack{\alpha\subseteq I_n\\ |\alpha|=m=1}} e_{\alpha,\alpha}$$

Proving our claim.

Using the case $m = 1$ and our claim, we get by induction on m that the spectrum of $A^*_{n,m} A_{n,m}$ consists of only one non-zero number which is $\lambda = \sum_{k=1}^n |a_k|^2$, and its multiplicity is $\binom{n-1}{m-1}$. Since each $x_{n,k,m}$ is a partial isometry of rank $\binom{n-1}{m-1}$, we get for $p = 1,\infty$

$$(2.31)\qquad \Big\|\sum_{k=1}^n a_k h_k\Big\|_p = \binom{n-1}{m-1}^{-1/p} \|A_{n,m}\|_p = \Big(\sum_{k=1}^n |a_k|^2\Big)^{1/2}.$$

Clearly, the orthogonal projection (2.24) onto $[h_k]_{k=1}^n$ is contractive since the computation (2.25) works also in this case. □

<u>Proposition 2.5:</u> Let $n < \infty$ and let $y_1,\dots,y_n \in C_1$ (respectively, $y_1,\dots,y_n \in C_\infty$) so that $\sum_{m=1}^n \|y_m\|_1 \binom{n-1}{m-1} = 1$ (respectively, $\max_{1\le m\le n} \|y_m\|_\infty = 1$). Put

(2.32) $$h^{(p)}_{n,k}(y_1,\dots,y_n) = \sum_{m=1}^{n} x_{n,k,m} \otimes y_m \;, \qquad p = 1,\infty$$

Then for any scalars $a_1,\dots,a_n$

(2.33) $$\Big\| \sum_{k=1}^{n} a_k h^{(p)}_{n,k}(y_1,\dots,y_n) \Big\|_p = \Big(\sum_{k=1}^{n} |a_k|^2 \Big)^{1/2}$$

and there is a contractive projection from C_p onto

$$H^n_p(y_1,\dots,y_n) = \mathrm{span}\{h^{(p)}_{n,k}(y_1,\dots,y_n)\}_{k=1}^{n}, \quad p = 1,\infty .$$

Proof: Let $p = 1$, then by proposition 2.4 and the remark preceeding the statement of the present proposition

(2.34) $$\Big\| \sum_{k=1}^{n} a_k h^{(1)}_{n,k}(y_1,\dots,y_n) \Big\|_1 = \sum_{m=1}^{n} \Big\| \sum_{k=1}^{n} a_k x_{n,k,m} \otimes y_m \Big\|_1 =$$

$$= \sum_{m=1}^{n} \|y_m\|_1 \cdot \Big\| \sum_{k=1}^{n} a_k x_{n,k,m} \Big\|_1 = \sum_{m=1}^{n} \|y_m\|_1 \cdot \Big(\sum_{k=1}^{n} |a_k|^2 \Big)^{1/2} \cdot \binom{n-1}{m-1} =$$

$$= \Big(\sum_{k=1}^{n} |a_k|^2 \Big)^{1/2} .$$

The proof of (2.33) for $p = \infty$ is the same, and it works even if we assume only that $y_j \in B(\ell_2)$.

As for contractive projection of C_p onto $H^n_p(y_1,\dots,y_n)$, if $p = 1$ we define for $x \in C_1$

(2.35) $$Px = \sum_{k=1}^{n} \langle x, h^{(\infty)}_{n,k}(v(y_1),\dots,v(y_n)) \rangle \, h^{(1)}_{n,k}(y_1,\dots,y_n)$$

and if $p = \infty$ we define for $x \in C_\infty$

(2.36) $$Px = \sum_{k=1}^{n} \langle x, h^{(1)}_{n,k}(z_1,\dots,z_n) \rangle \, h^{(\infty)}_{n,k}(y_1,\dots,y_n)$$

where $z_j \in C_1$ with $\|z_j\|_1 = 1 = \langle y_j, z_j \rangle$, $1 \le j \le n$.

The proof that the projections (2.35) and (2.36) are contractive (in C_1 and C_∞ respectively) is the same as (2.25), but now (2.21) is replaced by (2.23). □

In the same manner one can prove the following analogous of Proposition 2.5:

Proposition 2.6: Let $y_1, y_\infty \in C_1$ (respectively $y_1, y_\infty \in C_\infty$) so that $\|y_1\|_1 + \|y_\infty\|_1 = 1$ (respectively $\max\{\|y_1\|_\infty, \|y_\infty\|_\infty\} = 1$). Put for $p = 1,\infty$

$$(2.37) \qquad h^{(p)}_{\infty,k} = x_{\infty,k,1} \otimes y_1 + x_{\infty,k,\infty} \otimes y_\infty$$

Then for every scalars $\{a_k\}_{k=1}^{\infty}$ we have

$$(2.38) \qquad \left\| \sum_{k=1}^{\infty} a_k h^{(p)}_{\infty,k} \right\|_p = \left(\sum_{k=1}^{\infty} |a_k|^2 \right)^{1/2} ; \quad p = 1,\infty$$

and there is a contractive projection from C_p onto $H_p(y_1,y_\infty) = H_p^\infty(y_1,y_\infty) = \overline{\text{span}}\, \{h^{(p)}_{\infty,k}\}_{k=1}^{\infty}$, $p = 1,\infty$.

In this case the obvious analogous of the projections in formulas (2.35) and (2.36) are called "canonical" projections onto $H_1(y_1,y_\infty)$ and $H_\infty(y_1,y_\infty)$ respectively.

In the notations of propositions 2.5 and 2.6 we put:

Definition 2.7: Let $2 \leq n \leq \infty$. An elementary Hilbert subspace of C_p $(p = 1,\infty)$ is a space of the form $H_p^n(y_1,\dots,y_n)$ for $n < \infty$, or $H_p(y_1,y_\infty)$ if $n = \infty$. These elementary subspaces of C_p are said to be of type 4.

Note that $H_p^3(0,x,0)$ is identified naturally with $A_p^3(x)$ via their generating vectors: there exists unitary matrices w and u so that $\{w \cdot h^{(p)}_{3,k}(0,x,0) \cdot u\}_{k=1}^{3} = \{a^{(p)}_{i,j} \otimes x\}_{1\leq i<j\leq 3}$ where $a^{(p)}_{i,j} = a_{i,j}/\|a_{i,j}\|_p$ (these w and u correspond to the obvious permutations of rows and change of the sign in rows which transform any matrix

$\begin{pmatrix} -t_2 & t_1 & 0 \\ -t_3 & 0 & t_1 \\ 0 & -t_3 & t_2 \end{pmatrix}$ into $\begin{pmatrix} 0 & t_1 & t_2 \\ -t_1 & 0 & t_3 \\ -t_2 & -t_3 & 0 \end{pmatrix}$). Also, the one dimensional Hilbert space is classified as the elementary subspace of C_p of type 2, i.e. $SY_p^1(x)$.

c. Type 5: Elementary subspaces of C_p of the form $AH_p^n(y,z)$, $(p = 1,\infty)$

For $n < \infty$ and $1 \leq k \leq n$ we define

(2.39)
$$x_{n,k} = \sum_{\substack{k\in\alpha\subset I_n \\ |\alpha| \text{odd}}} (-1)^{i(\alpha,k)} \cdot e_{\alpha,\alpha \sim \{k\}} = \sum_{\substack{m \text{ odd} \\ 1\leq m\leq n}} x_{n,k,m}$$

$$\tilde{x}_{n,k} = \sum_{\substack{k\in\alpha\subset I_n \\ |\alpha| \text{ even}}} (-1)^{i(\alpha,k)} e_{\alpha \sim \{k\},\alpha}$$

and put $AH_p^n = \text{span}\,\{x_{n,k};\ \tilde{x}_{n,k}\}_{k=1}^n$, the "p" indicates that we regard this space as a subspace of C_p, $(p = 1,\infty)$.

Definition 2.8: Let $y,z \in C_p$ $(p = 1,\infty)$ with $y \perp z$ and $\|y + z\|_p = 1$. We denote for $2 \leq n < \infty$

$$(2.40) \qquad AH_p^n(y,z) = \text{span}\{x_{n,k} \otimes y + (\tilde{x}_{n,k})^T \otimes z,\ \tilde{x}_{n,k} \otimes y + (x_{n,k})^T \otimes z\}_{k=1}^n$$

subspaces of C_p of form $AH_p^n(y,z)$ are called elementary subspaces of type 5.

It can be easily verified that each of the $x_{n,k}$, $\tilde{x}_{n,k}$ is a partial isometry of rank 2^{n-2} and that by our treatment of elementary Hilbert subspaces, we have for every scalars $a_1,\ldots,a_n$:

$$(2.41) \qquad \left\|\sum_{k=1}^n a_k x_{n,k}\right\|_p = \left\|\sum_{k=1}^n a_k \tilde{x}_{n,k}\right\|_p = 2^{\frac{n-2}{p}} \left(\sum_{k=1}^n |a_k|^2\right)^{1/2}$$

Thus, AH_p^n is the span of the two elementary Hilbert subspaces $[x_{n,k}]_{k=1}^n$ and $[\tilde{x}_{n,k}]_{k=1}^n$ which are in a sense the adjoint of each other. This motivates our notation AH_p^n ("A" for adjoint and "H" for Hilbert). Note also that AH_p^2 is just $C_p^{2,2}$ and AH_p^3 is just A_p^4. Indeed, it is very easy to show that $\sum_{k=1}^n (a_k\, x_{n,k} + b_k\, \tilde{x}_{n,k})$ is unitarily equivalent to

$$\begin{pmatrix} a_1 & -b_2 \\ a_2 & b_1 \end{pmatrix} \quad \text{if} \quad n = 2, \text{ and to } \begin{pmatrix} 0 & a_1 & a_2 & a_3 \\ -a_1 & 0 & b_3 & -b_2 \\ -a_2 & -b_3 & 0 & b_1 \\ -a_3 & b_2 & -b_1 & 0 \end{pmatrix} \quad \text{if} \quad n = 3.$$

Thus, up to the change of the sign of $\tilde{x}_{2,2}$ and $\tilde{x}_{3,2}$, the vectors $\{x_{2,k}, \tilde{x}_{2,k}\}_{k=1}^2$ and $\{x_{3,k}, \tilde{x}_{3,k}\}_{k=1}^3$ correspond naturally to the standard unit matrices of $C_p^{2,2}$ and to the elementary anti-symmetric matrices of A_p^4 respectively. One more example is

$\sum_{k=1}^4 (a_k x_{4,k} + b_k \tilde{x}_{4,k}) =$

a_1					$-b_2$	$-b_3$	$-b_4$								
a_2					b_1			$-b_3$	$-b_4$						
a_2						b_1		b_2		$-b_4$					
a_4							b_1		b_2	b_3					
					a_3	$-a_2$		a_1							$-b_4$
					a_4		$-a_2$		a_1						b_3
						a_4	$-a_3$			a_1					$-b_2$
								a_4	$-a_3$	a_2					b_1

Another preliminary observation which follows directly from the definition is that for every scalars $\{a_k\}_{k=1}^n$ and $\{b_k\}_{k=1}^n$

$$(2.42)\qquad \Big\|\sum_{k=1}^n (a_k x_{n,k} + b_k \widetilde{x}_{n,k})\Big\|_p = \Big\|\sum_{k=1}^n (a_k \widetilde{x}_{n,k} + b_k x_{n,k})\Big\|_p, \quad (p = 1,\infty)$$

Proposition 2.9: Let $y,z \in C_p$ $(p = 1,\infty)$ with $y \perp z$ and $\|y+z\|_p = 1$. Then the mapping $x_{n,k} \to x_{n,k} \otimes y + (\widetilde{x}_{n,k})^T \otimes z$, $\widetilde{x}_{n,k} \to \widetilde{x}_{n,k} \otimes y + (x_{n,k})^T \otimes z$ extends to a linear isometry of AH_p^n onto $AH_p^n(y,z)$, and there is a canonical contractive projection from C_p onto $AH_p^n(y,z)$, $(p = 1,\infty)$.

Proof: Let $\{a_k\}_{k=1}^n$ and $\{b_k\}_{k=1}^n$ be scalars, then:

$$(2.43)\qquad \Big\|\sum_{k=1}^n a_k(x_{n,k} \otimes y + \tilde{x}_{n,k}^T \otimes z) + \sum_{k=1}^n b_k(\tilde{x}_{n,k} \otimes y + x_{n,k}^T \otimes z)\Big\|_1$$

$$= \Big\|\sum_{k=1}^n (a_k x_{n,k} + b_k \tilde{x}_{n,k}) \otimes y + \sum_{k=1}^n (a_k \tilde{x}_{n,k} + b_k x_{n,k})^T \otimes z\Big\|_1$$

$$= \Big\|\sum_{k=1}^n (a_k x_{n,k} + b_k \tilde{x}_{n,k})\Big\|_1 \|y\|_1 + \Big\|\sum_{k=1}^n a_k \tilde{x}_{n,k} + b_k x_{n,k}\Big\|_1 \|z\|_1 =$$

$$= \Big\|\sum_{k=1}^n a_k x_{n,k} + b_k \tilde{x}_{n,k}\Big\|_1 .$$

This establishes our first statement for $p = 1$; the proof for $p = \infty$ is identical it works even for $y,z \in B(\ell_2)$.

Now, the proof of the existence of a contractive projection from C_p onto $AH_p^n(y,z)$ reduces to the proof that the following canonical projection is contractive in C_p $(p = 1,\infty)$:

$$(2.44)\qquad P_n x = \sum_{k=1}^n \langle x, \frac{x_{n,k}}{2^{n-2}}\rangle\, x_{n,k} + \sum_{k=1}^n \langle x, \frac{\tilde{x}_{n,k}}{2^{n-2}}\rangle\, \tilde{x}_{n,k} .$$

Indeed, if P_n is contractive in C_∞ say, we define in C_1

$$(2.45)\quad Qx = \sum_{k=1}^{n} \langle x, x_{n,k} \otimes v(y) + \tilde{x}^T_{n,k} \otimes v(z)\rangle (x_{n,k} \otimes y + \tilde{x}_{n,k} \otimes z)/2^{n-2}$$

$$+ \sum_{k=1}^{n} \langle x, \tilde{x}_{n,k} \otimes v(y) + x^T_{n,k} \otimes v(z)\rangle (\tilde{x}_{n,k} \otimes y + x^T_{n,k} \otimes z)/2^{n-2}$$

and then

$$(2.46)\quad \|Qx\|_1 = \Big\| \sum_{k=1}^{n} \langle x, x_{n,k} \otimes v(y) + \tilde{x}^T_{n,k} \otimes v(z)\rangle x_{n,k}/2^{n-2}$$

$$+ \sum_{k=1}^{n} \langle x, \tilde{x}_{n,k} \otimes v(y) + x^T_{n,k} \otimes v(z)\rangle \tilde{x}_{n,k}/2^{n-2}\Big\|_1$$

$$= \sup_{\substack{a\in C_\infty \\ \|a\|_\infty \le 1}} \Big|\langle x, \sum_{k=1}^{n} \langle a, x_{n,k}/2^{n-2}\rangle (x_{n,k} \otimes v(y) + \tilde{x}^T_{n,k} \otimes v(z))$$

$$+ \sum_{k=1}^{n} \langle a, \tilde{x}_{n,k}/2^{n-2}\rangle (\tilde{x}_{n,k} \otimes v(y) + x^T_{n,k} \otimes v(z))\rangle\Big|$$

$$\le \|x\|_1 \sup_{\substack{a\in C_\infty \\ \|a\|_\infty \le 1}} \Big\| \sum_{k=1}^{n} \langle a, x_{n,k}/2^{n-2}\rangle x_{n,k}$$

$$+ \sum_{k=1}^{n} \langle a, \tilde{x}_{n,k}/2^{n-2}\rangle \tilde{x}_{n,k}\Big\|_\infty$$

$$= \|x\|_1 \cdot \sup_{\substack{a\in C_\infty \\ \|a\|_\infty \le 1}} \|P_n a\|_\infty = \|x\|_1 .$$

In a very similar manner one can show that the contracitivity of P_n in C_1 implies the existence of a contractive projection of C_∞ onto $AH^n_\infty(y,z)$.

The contractivity of the projection (2.44) will be proved by induction on n. For $n = 2$, P_2 is just the canonical projection of C_p onto $C_p^{2,2}$ which is contractive. Assume that $n \geq 3$ and that P_{n-1} is contractive for $p = 1,\infty$. The contractiveness of P_n will follow from that of P_{n-1} and the following.

<u>Lemma 2.10:</u> Let $a_{n-1}, a_n, b_{n-1}, b_n$ be scalars. Then there exist unitary operators v_1, v_2 so that for $1 \leq k \leq n-2$:

$$(2.47)\quad v_1 x_{n,k} v_2 = x_{n-1,k} + \tilde{x}^T_{n-1,k} ; \quad v_1 \tilde{x}_{n,k} v_2 = \tilde{x}_{n-1,k} + x^T_{n-1,k}$$

$$(2.48)\quad y = v_1^*(x_{n-1;n-1} + \tilde{x}^T_{n-1,n-1})v_2^* \text{ and } \tilde{y} = v_1^*(\tilde{x}_{n-1,n-1} + x^T_{n-1,n-1})v_2^*$$

belong to span $\{x_{n,n-1}, x_{n,n}, \tilde{x}_{n,n-1}, \tilde{x}_{n,n}\}$

$$(2.49)\quad v_1(a_{n-1} x_{n,n-1} + a_n x_{n,n} + b_{n-1} \tilde{x}_{n,n-1} + b_n \tilde{x}_{n,n}) v_2 =$$

$$= s(x_{n-1,n-1} + \tilde{x}^T_{n-1,n-1)} + t(\tilde{x}_{n-1,n-1} + x^T_{n-1,n-1})$$

where $|s|$ and $|t|$ are the s-numbers of the matrix $\begin{pmatrix} a_{n-1} & -b_n \\ a_n & b_{n-1} \end{pmatrix}$.

Assuming the lemma, let us continue the proof of proposition 2.9. Let $x \in C_p$ and let $z = P_n x = \sum_{k=1}^{n} (a_k x_{n,k} + b_k \tilde{x}_{n,k})$. Let v_1, $v_2, y, \tilde{y}, s$, and

t be as in lemma 2.10, and let Q be the canonical projection onto the span of $\{x_{n-1,k} + \tilde{x}^T_{n-1,k}, \tilde{x}_{n-1,k} + x^T_{n-1,k}\}_{k=1}^{n-1}$. Q is contractive since P_{n-1} is contractive. Thus $\tilde{Q}a = v_1^* \cdot Q(v_1\ a\ v_2) \cdot v_2^*$ defines the canonical contractive projection onto $\text{span}(\{x_{n,k}, \tilde{x}_{n,k}\}_{k=1}^{n-2} \cup \{y,\tilde{y}\})$. By (2.48) $P_n\, y = y$ and thus using (2.49) we get $s = \langle z,y\rangle/2^{n-2} = \langle P_n x,y\rangle/2^{n-2} =$ $= \langle x,y\rangle/2^{n-2}$, and similarly $t = \langle z,\tilde{y}\rangle/2^{n-2} = \langle x,\tilde{y}\rangle/2^{n-2}$. Thus using (2.49) we get:

$$(2.50)\quad \tilde{Q}x = \sum_{k=1}^{n-2} \left(\langle x, \frac{x_{n,k}}{2^{n-2}}\rangle x_{n,k} + \langle x, \frac{\tilde{x}_{n,k}}{2^{n-2}}\rangle \tilde{x}_{n,k}\right) + \langle x, \frac{y}{2^{n-2}}\rangle y + \langle x, \frac{\tilde{y}}{2^{n-2}}\rangle \tilde{y}$$

$$= \sum_{k=1}^{n-2} (a_k\, x_{n,k} + b_k\, \tilde{x}_{n,k}) + sy + t\tilde{y}$$

$$= \sum_{k=1}^{n} (a_k\, x_{n,k} + b_k\, \tilde{x}_{n,k}) = P_n x.$$

So, $\|P_n x\|_p = \|\tilde{Q}x\|_p \leq \|x\|_p$ by the contractivity of $\tilde{Q}$.

<u>Proof of lemma 2.10</u>

We choose first untary 2×2 matrices $w = \begin{pmatrix} w_{1,1} & w_{1,2} \\ w_{2,1} & w_{2,2} \end{pmatrix}$ and $u = \begin{pmatrix} u_{1,1} & u_{1,2} \\ u_{2,1} & u_{2,2} \end{pmatrix}$ with $\det w = \det u = 1$ and

$$(2.51)\quad \begin{pmatrix} w_{1,1} & w_{1,2} \\ w_{2,1} & w_{2,2} \end{pmatrix} \begin{pmatrix} a_{n-1} & -b_n \\ a_n & b_{n-1} \end{pmatrix} \begin{pmatrix} u_{1,1} & u_{1,2} \\ u_{2,1} & u_{2,2} \end{pmatrix} = \begin{pmatrix} 0 & s \\ -t & 0 \end{pmatrix}.$$

Hence,

$$(2.52)\quad \begin{pmatrix} -u_{2,1} & u_{1,1} \\ u_{2,2} & -u_{1,2} \end{pmatrix} \begin{pmatrix} -b_{n-1} & -b_n \\ a_n & -a_{n-1} \end{pmatrix} \begin{pmatrix} -w_{1,2} & w_{2,2} \\ w_{1,1} & -w_{2,1} \end{pmatrix} = \begin{pmatrix} 0 & -t \\ s & 0 \end{pmatrix}.$$

Let $a_1,\ldots,a_{n-2}$ and $b_1,\ldots,b_{n-2}$ be scalars. For every $1 \leq m \leq n$ we define matrices $0_m = 0_m(a_1,\ldots,a_m;b_1,\ldots,b_m)$ and

$E_m = E_m(a_1,\dots,a_m;\ b_1,\dots,b_m)$ by:

$$(2.53)\quad O_m = \sum_{k=1}^{m} (a_k\, x_{m,k} + b_k\, \tilde{x}_{m,k});\quad E_m = \sum_{k=1}^{m} (a_k\, \tilde{x}^T_{m,k} + b_k\, x^T_{m,k})$$

By the definition of the $x_{m,k}$ and $\tilde{x}_{m,k}$ we have:

$$(2.54)\quad O_n = O_{n-1} + E^{(n)}_{n-1} + a_n\, x_{n,n} + b_n\, \tilde{x}_{n,n} =$$

$$= O_{n-2} + E^{(n-1)}_{n-2} + E^{(n)}_{n-2} + O^{(\{n-1,n\})}_{n-2} + a_{n-1}\, x_{n,n-1} + a_n x_{n,n} +$$

$$+ b_{n-1}\, \tilde{x}_{n,n-1} + b_n \tilde{x}_{n,n}$$

So:

$$(2.55)\quad O_n = \left[\begin{array}{|c|c|c|c|c|c|c|c|} \hline & & & & O_{n-2} & -b_{n-1}I & -b_nI & \\ \hline & & & & a_{n-1}I & E^{(n-1)}_{n-2} & & -b_nI \\ \hline & & & & a_nI & & E^{(n)}_{n-2} & b_{n-1}I \\ \hline & & & & & a_nI & -a_{n-1}I & O^{(\{n-1,n\})}_{n-2} \\ \hline & & & & & & & \\ \hline & & & & & & & \\ \hline & & & & & & & \\ \hline & & & & & & & \\ \hline \end{array}\right]$$

Let $\tilde{v}_1$ and $\tilde{v}_2$ be the following partial isometries:

$$(2.56)\qquad \tilde{v}_1 = \left[\begin{array}{|c|c|c|c|c|c|c|c|}
\hline
-u_{2,1}I & & & u_{1,1}I & & & & \\ \hline
 & w_{1,1}I & w_{1,2}I & & & & & \\ \hline
 & & & & & & & \\ \hline
 & & & & & & & \\ \hline
u_{2,2}I & & & -u_{1,2}I & & & & \\ \hline
 & w_{2,1}I & w_{2,2}I & & & & & \\ \hline
 & & & & & & & \\ \hline
 & & & & & & & \\ \hline
\end{array}\right]$$

$$(2.57)\qquad \tilde{v}_2 = \left[\begin{array}{|c|c|c|c|c|c|c|c|}
\hline
 & & & & & & & \\ \hline
 & & & & & & & \\ \hline
 & & & & & & & \\ \hline
 & & & & & & & \\ \hline
u_{1,1}I & & & & u_{1,2}I & & & \\ \hline
 & -w_{1,2}I & & & & w_{2,2}I & & \\ \hline
 & w_{1,1}I & & & & -w_{2,1}I & & \\ \hline
u_{2,1}I & & & & u_{2,2}I & & & \\ \hline
\end{array}\right]$$

Using (2.51), (2.52) and $\det w = \det u = 1$, we get:

(2.58) $\tilde{v}_1 \cdot 0_n \cdot \tilde{v}_2 =$

				0_{n-2}	$-tI$		
				sI	$E^{(n-1)}_{n-2}$		
E_{n-2}	sI						
$-tI$	$0^{(n-1)}_{n-2}$						

$$= 0_{n-1}(a_1,\ldots,a_{n-2},s;b_1,\ldots,b_{n-2},t) + E_{n-1}(a_1,\ldots,a_{n-2},s;b_1,\ldots,b_{n-2},t)$$

Let us choose some unitary extensions v_1 and v_2 of $\tilde{v}_1$ and $\tilde{v}_2$ respectively. Then $v_1 0_n v_2 = 0_{n-1}(a_1,\ldots,a_{n-2},s;b_1,\ldots,b_{n-2},t) + E_{n-1}(a_1,\ldots,a_{n-2},s;b_1,\ldots,b_{n-2},t)$, so by the definition of the matrices of form 0_m and E_m and by the linear independence of the system $\{x_{m,k}, \tilde{x}_{m,k}\}_{k=1}^m$ for $m = n$ and $m = n-1$, we get (2.47) and (2.49). Finally, (2.48) follows easily from the construction of v_1 and v_2. □

<u>Remark 2.11:</u> Lemma 2.10 gives us an algorithm to compute the s-numbers of a matrix $0_n = \sum_{k=1}^{n} (a_k x_{n,k} + b_k \tilde{x}_{n,k})$, and thus - its norm. Indeed, let $t_n^{(1)} = a_n$, $t_n^{(2)} = b_n$. For $1 \leq m \leq n-1$ we construct inductively unitary matrices $v_{n-m}^{(1)}$ and $v_{n-m}^{(2)}$ and numbers $t_{n-m}^{(1)}$, $t_{n-m}^{(2)}$ so that $|t_{n-m}^{(1)}|$ and $|t_{n-m}^{(2)}|$ are the two s-numbers of the matrix

$$(2.59) \qquad \begin{pmatrix} a_{n-m} & -t_{n-m+1}^{(2)} \\ t_{n-m+1}^{(1)} & b_{n-m} \end{pmatrix}$$

and if we put $\tilde{O}_{n-m} = O_{n-m}(a_1,\dots,a_{n-m-1},t^{(1)}_{n-m};\ b_1,\dots,b_{n-m-1},t^{(2)}_{n-m})$ and

$\tilde{E}_{n-m} = E_{n-m}(a_1,\dots,a_{n-m-1},\ t^{(1)}_{n-m};\ b_1,\dots,b_{n-m-1},t^{(2)}_{n-m})$, we get:

$$(2.60)\quad v^{(1)}_{n-m}\cdot O_n\cdot v^{(2)}_{n-m} = \sum_{\substack{\alpha\subseteq\{n-m+1,\dots,n\}\\ |\alpha|\ \text{even}}} \tilde{O}^{(\alpha)}_{n-m} + \sum_{\substack{\alpha\subseteq\{n-m+1,\dots,n\}\\ |\alpha|\ \text{odd}}} \tilde{E}^{(\alpha)}_{n-m}\ .$$

Note that in this sum the summands are pairwise disjointly supported. It follows by putting $m=n-1$ in (2.60) that there exist unitary matrices $u^{(1)}$ and $u^{(2)}$ so that:

$$(2.61)\quad u^{(1)}\cdot O_n\cdot u^{(2)} = \operatorname{diag}\{\overbrace{t^{(1)}_1,\dots,t^{(1)}_1}^{2^{n-2}\ \text{terms}};\ \overbrace{t^{(2)}_1,\dots,t^{(2)}_1}^{2^{n-2}\ \text{terms}}\}\ .$$

In particular, the s-numbers of O_n are $|t^{(1)}_1|$ and $|t^{(2)}_1|$, each with multiplicity 2^{n-2}.

d. <u>Type 6: Elementary subspaces of C_p of the form $DAH^n_p(x)$ $(p = 1,\infty)$</u>

Let $n < \infty$ and let the vectors $x_{n,k}$ and $\tilde{x}_{n,k}$ be defined by formulas (2.39), $1 \leq k \leq n$. Let

$$(2.62)\quad y_n = x_{n,n} + \tilde{x}_{n,n}$$

and denote $DAH^n_p = \operatorname{span}(\{x_{n,k},\tilde{x}_{n,k}\}^{n-1}_{k=1} \cup \{y_n\})$. As before, the "p" indicates that we regard this space as a subspace of C_p, $p = 1,\infty$. By our treatment of AH^n_p it is clear that $\{x_{n,k},\tilde{x}_{n,k}\}^{n-1}_{k=1}$ are isometrically equivalent in the natural ordering to $\{2^{1/p}x_{n-1,k}, 2^{1/p}\tilde{x}_{n-1,k}\}^{n-1}_{k=1}$. Thus, DAH^n_p contains AH^{n-1}_p and is contained in AH^n_p. Note that DAH^1_p is one dimensional, and that DAH^2_p is identified in the obvious way with SY^2_p.

<u>Definition 2.12:</u> Let $x \in C_p$ $(p = 1,\infty)$, $\|x\|_p = 1$, and let $3 \leq n < \infty$. We define

$$(2.63)\quad DAH^n_p(x) = \operatorname{span}(\{x_{n,k} \otimes x,\ \tilde{x}_{n,k} \otimes x\}_{k=1}^{n-1} \cup \{y_n \otimes x\}),$$

We call subspaces of C_p of the form $DAH^n_p(x)$ elementary subspaces of type 6.

Proposition 2.13: Let $x \in C_p$, $\|x\|_p = 1$, $p = 1,\infty$ and let $3 \leq n < \infty$. Then the mapping $a \to a \otimes x$ is an isometry of DAH^n_p onto $DAH^n_p(x)$, and there is a canonical contractive projection of C_p onto $DAH^n_p(x)$.

Proof: The first statement is obvious. Let us define for $a \in C_p$:

$$(2.64)\quad R_n a = \sum_{k=1}^{n-1} (\langle a, \frac{x_{n,k}}{2^{n-2}} \rangle x_{n,k} + \langle a, \frac{\tilde{x}_{n,k}}{2^{n-2}} \rangle \tilde{x}_{n,k}) + \langle a, \frac{y_n}{2^{n-1}} \rangle y_n$$

We claim that this canonical projection of C_p onto DAH^n_p is contractive. Since the canonical projection (2.44) of C_p onto AH^n_p is contractive, it is enough to show that $R_n|_{AH^n_p}$ is contractive. This however is a consequence of the following

Claim: the mapping on AH^n_p defined by $x_{n,n} \to \tilde{x}_{n,n}$ $\tilde{x}_{n,n} \to x_{n,n}$ and $x_{n,k} \to x_{n,k}$, $\tilde{x}_{n,k} \to \tilde{x}_{n,k}$ for $1 \leq k < n-1$ is an isometry of AH^n_p.

Indeed, let $\{a_k\}_{k=1}^n$, $\{b_k\}_{k=1}^n$ be scalars. If we interchange a_n and b_n in the proof of lemma 2.10, and replace w and u by $\begin{pmatrix} 0 & 1 \\ 1 & 0 \end{pmatrix} u^T \begin{pmatrix} 1 & 0 \\ 0 & -1 \end{pmatrix}$ and $\begin{pmatrix} 1 & 0 \\ 0 & -1 \end{pmatrix} w^T \begin{pmatrix} 0 & 1 \\ 1 & 0 \end{pmatrix}$ respectively, we get instead of $\tilde{v}_1$ and $\tilde{v}_2$ other partial isometries $\approx{v}_1$ and $\approx{v}_2$ respectively. Now it is easy to see that

$$(2.65)\quad \overset{\approx}{v}_1 \cdot O_n(a_1,\dots,a_{n-1},b_n;b_1,\dots,b_{n-1},a_n) \cdot \overset{\approx}{v}_2 = \tilde{v}_1 \cdot O_n(a_1,\dots,a_n;b_1,\dots,b_n) \cdot \tilde{v}_2$$

and thus $O_n(a_1,\dots,a_{n-1},b_n;b_1,\dots,b_{n-1},a_n)$ and $O_n(a_1,\dots,a_n;b_1,\dots,b_n)$ have the same s-numbers and thus - the same norm.

Now, as in the proof of proposition 2.9, the contractivity of R_n implies that the canonical projection of C_p onto $DAH^n_p(x)$ is contractive.

□

e. Formulation of the main results

After this lengthy description of elementary subspaces of C_p $(p = 1,\infty)$ we are ready to state our main results. We just remark that in our definitions of elementary subspaces of C_p, the orthonormal bases $\{e_i\}_{i=1}^\infty$ and $\{f_i\}_{i=1}^\infty$ of ℓ_2, and the tensor product representation of C_p as $C_p \otimes C_p$ using the bases $(\{e_i\}_{i=1}^\infty, \{f_i\}_{i=1}^\infty)$, were arbitrary. Thus if X is an elementary subspace of C_p and if u and w are partial isometries with $r(u) \geq \ell(X)$ and $r(w) \geq r(X)$, then $u X w^*$ is an elementary subspace of C_p. The same is also trivially true for $X \otimes x_o$, where $x_o \in C_p$ and $\|x_o\|_p = 1$. Also, an elementary subspace X is called simple if it is of the simplest possible form. Precisely, the simple elementary subspaces of C_p are: $C_p^{n,m}$, SY_p^m, A_p^m, $\ell_2^n = [e_{k,1}]_{k=1}^n$, AH_p^n, DAH_p^n where $1 \leq n,m \leq \infty$ and $p = 1,\infty$.

Theorem 2.14: A subspace X of C_1 is the range of a contractive projection from C_1 if and only if $X = \sum_{k \in K} X_k$, $1 \leq |K| \leq \aleph_o$, where the X_k are pairwise disjointly supported subspaces of X, and each of the X_k is an elementary subspace of C_1. Consequently, X is isometric to the ℓ_1-sum of a sequence of simple elementary subspaces of C_1.

Theorem 2.15: Let P be a contractive projection in C_1, and let $R(P) = X = \sum_{k \in K} X_k$ be as in theorem 2.14. Let Q_k be the canonical projection of C_1 onto X_k. Then for each $a \in C_1$ the sum $Qa = \sum_{k \in K} Q_k a$ converges in C_1, and Q is a contractive projection of C_1 onto X, called the "canonical" projection. The operator $T = P - Q$ satisfies $\|T\| \leq 1$, $T^2 = 0$ and $T = PT = QT = T\,F(X)$.

Conversely, let Q be any such canonical contractive projection in C_1, let $R(Q) = X$, and let T be an operator on C_1 with $\|T\| \leq 1$, $T^2 = 0$ and $T = QT = T\,F(X)$. Then $P = Q + T$ is a contractive projection from C_1 onto X.

Theorem 2.16: Let P be a contractive projection in C_∞, and let $X = R(P)$. Then there exist a family $\{Y_k\}_{k \in K}$, $1 \leq |K| \leq \aleph_o$, of pairwise disjointly supported elementary subspaces of C_∞, a canonical contractive projection Q of C_∞ onto $Y = \sum_{k\in K} Y_k$, and an operator T on C_∞ with $\|T\| \leq 1$, $T^2 = 0$, and $T = TQ = F(Y)T$, so that $P = Q + T$. Also, $X = \sum_{k\in K} X_k$ where $X_k = (1 + T)Y_k$, and thus X is isometric to the c_0-sum of simple elementary subspaces of C_∞.

Conversely, let $\{Y_k\}_{k\in K}$, Q and T as above, then $P = Q + T$ is a contractive projection of C_∞ onto $R(Q) = X$.

That a contractive projection of C_p $(p = 1,\infty)$ onto its subspace X need not be unique, is shown in the following examples. For $p = 1$, we choose any $x \in C_1$, $\|x\|_1 = 1$ so that $v(x) \neq I$, and then for any $0 \neq y_0 \in B(\ell_2)$ with $\|y_0\|_\infty \leq 1$ and $y_0 \perp v(x)$ we define $Pa = \langle a, v(x) + y_0\rangle x$, $a \in C_1$. For $p = \infty$, we take any $x \in C_\infty$, $\|x\|_\infty = 1$, and we take any partial isometry v and $\tilde{x} \in C_\infty$ with $x = v + \tilde{x}$ and $v \perp \tilde{x}$. If $y_0 \in C_1$ is so that $1 = \|y_0\|_1 = \langle v, y_0\rangle$, we define $Pa = \langle a, y_0\rangle x$. It is clear that in both cases P is a contractive projection onto the one-dimensional subspace spanned by x, and the canonical projection Q is $a \to \langle a, v(x)\rangle x$ in the first case and $a \to \langle a, y_0\rangle v$ in the second case.

Note also that in theorem 2.15 the projection Q is the only contractive projection of C_1 onto X which satisfies $Q = Q\,E(X)$ and that Q is the unique contractive projection of C_1 onto X if and only if $F(X) = 0$.

Two natural questions concerning the representation $X = \sum_{k\in K} X_k$ of theorems 2.14 and 2.16 are:

(*) Is the representation $X = \sum_{k\in K} X_k$ of a subspace of C_p $(p = 1,\infty)$ onto which there is a contractive projection from C_p, unique?

(**) Let $\{Y_k\}_{k\in K}$ be a family of pairwise disjointly supported simple elementary subspaces of C_p $(p = 1,\infty)$, and let V be an isometry of $Y = \sum_{k\in K} Y_k$ into C_p. Does there exist a contractive projection from C_p onto $X = V(Y)$.

We shall deal with these two questions in section 6. Problem (*) is solved completely affirmatively in both cases $p = 1$ and $p = \infty$, while problem (**) is solved completely, almost affirmatively only for $p = 1$.

§3. THE RELATION BETWEEN A CONTRACTIVE PROJECTION P IN C_1 AND THE PROJECTIONS $E(x)$, $F(x)$ AND $G(x)$ FOR $x \in R(P)$

In this section we work in C_1. P will be some fixed contractive projection in C_1. Our first result shows the strong relations between P and the support projections $E(x)$, $F(x)$ and $G(x)$ for $x \in R(P)$. This is very similar to the result in the commutative case, i.e. in $L_1(\mu)$ spaces, where $E(x)$ and $F(x)$ are replaced by the operators of multiplication by the characteristic functions of the supports of x and its complement respectively.

Lemma 3.1: Let $0 \neq x \in R(P)$, and put $E = E(x)$, $F = F(x)$ and $G = G(x)$. Then:

(3.1) $PE = EPE$

(3.2) $EP = PEP$

(3.3) $FP = FPF = PFP$

(3.4) $GP = PGP$

In particular: PE, EP, FP and GP are contractive projections in C_1 (provided, of course that they are not zero).

Proof: We may assume that $\|x\|_1 = 1$. Let us choose a matrix representation in which every $y \in C_1$ is represented as a 2×2 operator matrix
$y = \begin{pmatrix} y_{1,1} & y_{1,2} \\ y_{2,1} & y_{2,2} \end{pmatrix}$, and such that

$$(3.5) \qquad x = \left(\begin{array}{cccc|c} s_1 & & & & \\ & \ddots & & & \\ & & s_i & & 0 \\ & & & \ddots & \\ \hline & & 0 & & 0 \end{array}\right)$$

where $\{s_i\}_{i\in I}$ is the sequence of all non zero s-numbers of x. The projections E, F and G acts in the following way on an element $y = \begin{pmatrix} y_{1,1} & y_{1,2} \\ y_{2,1} & y_{2,2} \end{pmatrix}$:

$$(3.6) \qquad Ey = \begin{pmatrix} y_{1,1} & 0 \\ 0 & 0 \end{pmatrix}, \quad Fy = \begin{pmatrix} 0 & 0 \\ 0 & y_{2,2} \end{pmatrix}, \quad GY = \begin{pmatrix} 0 & y_{1,2} \\ y_{2,1} & 0 \end{pmatrix}$$

We first establish the left equality of (3.3), which is clearly equivalent to

$$(3.7) \qquad FPG = 0, \quad FPE = 0 .$$

To show this, it is enough to show that $FPz = 0$ for every z of the following three forms in which z has only one non-zero coordinate:

$$(3.8) \quad z^{(1)} = \left(\begin{array}{c|ccc} & 0 & \overset{k}{|} & 0 \\ 0 & - & 1 & - \; i \\ & 0 & | & 0 \\ \hline 0 & & 0 & \end{array}\right), \quad z^{(2)} = \left(\begin{array}{ccc|c} \ddots & \overset{i}{|} & 0 & \\ j \; - & 1 & - & 0 \\ 0 & | & \ddots & \\ \hline & 0 & & 0 \end{array}\right), \quad z^{(3)} = \left(\begin{array}{ccc|c} \ddots & \overset{j}{|} & 0 & \\ i \; - & 1 & - & 0 \\ 0 & | & \ddots & \\ \hline & 0 & & 0 \end{array}\right)$$

Put $Pz^{(\ell)} = y^{(\ell)} = \begin{pmatrix} y^{(\ell)}_{1,1} & y^{(\ell)}_{1,2} \\ y^{(\ell)}_{2,1} & y^{(\ell)}_{2,2} \end{pmatrix}$, $\ell = 1,2,3$, and let t be any positive number. Then

$$(3.9) \quad 1 - s_i + (s_i^2 + t^2)^{1/2} = \|x + tz^{(1)}\|_1 \geq \|x + ty^{(1)}\|_1 \geq$$

$$\geq \|x + ty_{1,1}^{(1)}\|_1 + t\|y_{2,2}^{(1)}\|_1 \geq |< x + ty_{1,1}^{(1)}, v(x)>| + t\| y_{2,2}^{(1)}\|_1$$

$$= 1 + t \|y_{2,2}^{(1)}\|_1$$

since $P^* v(x) \in N(x)$ by Proposition 1.6, and since $< y_{1,1}^{(1)}, v(x)> =$ $=< y^{(1)}, v(x) > =< z^{(1)}, P^* v(x) > = 0$. Since t is arbitrary, (3.9) implies that $y_{2,2}^{(1)} = 0$. Also,

$$(3.10) \quad 1 + t = \|x + tz^{(2)}\|_1 \geq |< x + ty_{1,1}^{(2)}, v(x)>| + t \|y_{2,2}^{(2)}\|_1$$

$$= 1 + t + t \|y_{2,2}^{(2)}\|_1$$

since $< z^{(2)}, P^* v(x)> = 1$. Thus $y_{2,2}^{(2)} = 0$. Finally, since $< z^{(3)}, P^* v(x)> = 0$ we get that $y_{2,2}^{(3)} = 0$ from

$$(3.11) \quad 1 + t\|y_{2,2}^{(3)}\|_1 \leq \|x + tz^{(3)}\|_1 = 1 - (s_i + s_j) + \left\|\begin{pmatrix} s_i & 0 \\ t & s_j \end{pmatrix}\right\|_1 =$$

$$= 1 + t^2/2(s_i + s_j) + O(t^4)$$

This completessthe proof of (3.7).

From (3.7) we get $FP = FPFP$, and thus for every $y \in C_1$

$$(3.12) \quad \|FPy\|_1 = \|FPFPy\|_1 \leq \|PFPy\|_1 \leq \|FPy\|_1$$

and so equality holds everywhere. By Corollary 1.3 we get $FPy = PFPy$, proving the second equality in (3.3).

To prove (3.1), it is enough to show that $EPe_{i,j} = Pe_{i,j}$ for every $i,j \in I$. Now, since $x = \sum_{i\in I} s_i e_{i,i}$, we get

$$(3.13)\quad 1 = \|x\|_1 = \|EPx\|_1 \leq \sum_{i\in I} s_i \|EPe_{i,i}\|_1 \leq$$

$$\leq \sum_{i\in I} s_i\|Pe_{i,i}\|_1 \leq \sum_{i\in I} s_i = 1,$$

So equality holds everywhere and, again by Corollary 1.3, $Pe_{i,i} = EPe_{i,i}$ for every $i \in I$. If $i,j \in I$ and $i \neq j$ then $< e_{i,j}, P^* v(x) > = < e_{j,i}, P^* v(x) > = 0$. For every $\lambda \in C$ with $|\lambda| = 1$ we put $x_\lambda = e_{i,i} + \lambda e_{i,j} + \bar{\lambda} e_{j,i} + e_{j,j}$. Then:

$$(3.14)\quad 2 = \|x_\lambda\|_1 \geq \|Px_\lambda\|_1 \geq \|EPx_\lambda\|_1 \geq |<EPx_\lambda, v(x)>| = <x_\lambda, P^* v(x)> = 2.$$

By Corollary 1.3 we get for any such λ, $EPx_\lambda = Px_\lambda$, and from this it is now easy to conclude that $EPe_{i,j} = Pe_{i,j}$ for any $i,j \in I$, proving (3.1).

As for (3.4), from (3.1) and (3.3) we get

$$(3.15)\quad GP = GP(E + F + G) = GP(F + G) = GP(F + G)P = GPGP.$$

If $y \in C_1$ then

$$(3.16)\quad \|GPy\|_1 = \|GPGPy\|_1 \leq \|PGPy\|_1 \leq \|GPy\|_1,$$

and thus equality holds. Since by (3.3) $F(PGPy) = 0$, Proposition 1.4 implies that also $E(PGPy) = 0$ and thus $G(PGPy) = PGPy = GPy$, proving (3.4). Finally, by (3.3) and (3.4),

$$(3.17)\quad EP = EP(E + F + G)P = EPEP$$

and from this, (3.2) follows exactly as $FP = PFP$ follows from $FP = FPFP$

That PE, EP, FP and GP are projections is an easy consequence of (3.1), (3.2), (3.3) and (3.4). □

Note that the ranges of the projections (3.1), (3.2), (3.3) and (3.4) are the intersections of R(P) with R(E), R(E), R(F) and R(G) respectively. Also, the first example after theorem 2.16 shows that PF need not be a projection and EP need not be equal to PE.

In the study of contractive projections P in separable $L_1(\mu)$ spaces an important role is played by the sub-σ-algebra of sets which are the supports of elements of R(P): there is a function $f \in L_1(\mu)$ so that R(P) is generated by the functions of the form $f \cdot \chi_A$, where A is a support of some member of R(P). When the measure μ is purely atomic (i.e. $L_1(\mu) = \ell_1$), there is a sequence $\{A_j\}_{j \in I}$ of pairwise disjoint subsets of the positive integers and elements $x_j \in R(P)$ with $\operatorname{supp} x_j = A_j$ so that $R(P) = [x_j]_{j \in I} = \ell_1^n$ $(n = |I|)$. In a sense, the same situation is true also in C_1. Let us introduce the following definition:

<u>Definition 3.2:</u> An element $x \in R(P)$, $\|x\|_1 = 1$, is an <u>atom</u> of R(P) if x has minimal support in the following sense: if $0 \neq y \in R(E(x)P)$ then $E(x) = E(y)$.

For convenience we regard an atom of R(P) also as an atom of P, no confusion can arise since P determines R(P).

<u>Proposition 3.3:</u> Let $x \in R(P)$ with $\|x\|_1 = 1$. Then the following are equivalent:

(i) x is an atom of R(P);

(ii) x is an extreme point of the unit ball of R(P);

(iii) for every $y \in C_1$, $PE(x)y = \langle y, v(x)\rangle x$;

(iv) for every $y \in C_1$, $E(x)Py = \langle y, P^* v(x)\rangle x$.

Consequently, $R(P)$ has atoms.

<u>Proof:</u> The equivalence of (iv) and (iii) is an easy consequence of Propositions 1.5 and 1.6. If (iii) holds and if $y,z \in R(P)$, $\|y\|_1 = \|z\|_1 = 1$ and $x = (y+z)/2$, then:

$$(3.18)\qquad 1 = \|x\|_1 = \|E(x)x\|_1 = \|(E(x)y + E(x)z)/2\|_1 \leq$$

$$\leq (\|E(x)y\|_1 + \|E(x)z\|_1)/2 \leq 1$$

Thus, we have equality everywhere, and by Corollary 1.3 we get $E(x)y = y$, $E(x)z = z$. So, by (iii), we have

$$(3.19)\qquad y = \langle y, v(x)\rangle x \quad , \qquad z = \langle z, v(x)\rangle x$$

and since $x = (y+z)/2$ we get that $y = z = x$. So x is an extreme point of the unit ball of $R(P)$, and (iii) implies (ii).

Conversely, assume that x is an extreme point of the unit ball of $R(P)$. Choose a matrix representation in which $x = \sum_{i\in I} s_i e_{i,i}$ where $\{s_i\}_{i\in I}$ are all the non-zero s-numbers of x. Then $x = Px = \sum_{i\in I} s_i Pe_{i,i}$ is a convex combination of elements of $R(P)$ with norm ≤ 1, and so we must have $Pe_{i,i} = x$ for every $i \in I$. If $i,j \in I$ with $i \neq j$, and if $\lambda \in C$ with $|\lambda| = 1$, then:

$$(3.20)\qquad \|x \pm P(\lambda e_{i,j} + \bar{\lambda} e_{j,i})/2\|_1 = \|(P(e_{i,i} \pm \lambda e_{i,j} \pm \bar{\lambda} e_{j,i} + e_{j,j})/2\|_1$$

$$\leq \|(e_{i,i} \pm \lambda e_{i,j} \pm \bar{\lambda} e_{j,i} + e_{j,j})/2\|_1 = 1$$

So $P(\lambda e_{i,j} + \overline{\lambda} e_{j,i}) = 0$, and thus $Pe_{i,j} = Pe_{j,i} = Pe_{j,i} = 0$. If $y \in C_1$, write $E(x)y = \sum_{ij \in I} y_{i,j} e_{i,j}$, and then

$$\text{(3.21)} \qquad PE(x)y = \sum_{i,j \in I} y_{i,j} Pe_{i,j} = (\sum_{i \in I} y_{i,i})x = \langle y, v(x) \rangle x$$

proving (iii).

Let x be an atom of $R(P)$ and let y be an extreme point of the unit ball of $R(PE(x))$, which exists by Proposition 1.7. Since x is an atom, we have $E(x) = E(y)$, and using the equivalence (ii) $\Leftrightarrow$ (iii) we get that $x = PE(y)x = \lambda y$ for some λ with $|\lambda| = 1$. Hence x is an extreme point of the unit ball of $R(PE(x))$ which implies (using again the equivalence (ii) $\Leftrightarrow$ (iii) that x is an extreme point of the unit ball of $R(P)$. So (i) implies (ii). Finally, (i) is a trivial consequence of (iii).

A useful consequence of Proposition 3.3 is

Corollary 3.4: Let x be an atom of P, let $e \in \ell_2$ with $r(x)e = e$ and $\|e\| = 1$. Then $P((\cdot,e)v(x)e) = x$.

We conclude this section by a result which allows us to "translate" elements of $R(P)$, i.e. to conclude from $x \in R(P)$ that an appropriate "translation" of x belongs also to $R(P)$. We begin with an elementary example:

Example 3.5: Define projections $Q_1 \colon C_1^2 \to C_1^2$ and $Q_2 \colon C_1^3 \to C_1^3$ by

$$(3.22)\quad Q_1\begin{pmatrix} t_{1,1} & t_{1,2} \\ t_{2,1} & t_{2,2} \end{pmatrix} = \begin{pmatrix} t_{1,1} & t_{1,2} \\ t_{2,1} & 0 \end{pmatrix}$$

and

$$(3.23)\quad Q_2\begin{pmatrix} t_{1,1} & t_{1,2} & t_{1,3} \\ t_{2,1} & t_{2,2} & t_{2,3} \\ t_{3,1} & t_{3,2} & t_{3,3} \end{pmatrix} = \begin{pmatrix} t_{1,1} & (t_{1,2}+t_{2,1})/2 & (t_{1,3}+t_{3,1})/2 \\ (t_{1,2}+t_{2,1})/2 & t_{2,2} & 0 \\ (t_{1,3}+t_{3,1})/2 & 0 & t_{3,3} \end{pmatrix}$$

Then, Q_1 and Q_2 are not contractive, even when restricted to the symmetric matrices.

Indeed, $2 = \left\|\begin{pmatrix} 1 & 1 \\ 1 & 1 \end{pmatrix}\right\|_1$ and $\left\|\begin{pmatrix} 1 & 1 \\ 1 & 0 \end{pmatrix}\right\|_1 = \sqrt{5} > 2$, so Q_1 is not contractive. Also,

$$3 = \left\|\begin{pmatrix} 1 & 1 & 1 \\ 1 & 1 & 1 \\ 1 & 1 & 1 \end{pmatrix}\right\|_1, \text{ but } \left\|\begin{pmatrix} 1 & 1 & 1 \\ 1 & 1 & 0 \\ 1 & 0 & 1 \end{pmatrix}\right\|_1 > 3$$

Since the determinant of $A = \begin{pmatrix} 1 & 1 & 1 \\ 1 & 1 & 0 \\ 0 & 0 & 1 \end{pmatrix}$ is 1, and 1 is a simple eigenvalue of A. Thus, Q_2 is not contractive.

These simple examples are special cases of the following

Proposition 3.6: Let $a \in C_1$ with $\|a\|_1 = 1$, and let $s_{i,j}$ denote the elementary symmetric matrices.

(i) if $y_{1,1} = s_{1,1} \otimes a$, $y_{1,2} = s_{1,2} \otimes a \in R(P)$, then $y_{2,2} = s_{2,2} \otimes a \in R(P))$.

(ii) if $y_{i,j} = s_{i,j} \otimes a \in R(P)$ for all pairs (i,j) with $1 \leq i \leq j \leq 3$, except perhaps the pair $(2,3)$, then $y_{2,3} \in R(P)$.

Proof: (i) Since $E(y_{1,2})\, y_{2,2} = y_{2,2}$, we get by Lemma 3.1, Proposition 1.5 and Proposition 1.6 that

$$(3.25) \qquad E(y_{1,2})\, Py_{2,2} = Py_{2,2}, \quad \text{and} \quad \langle Py_{2,2}, v(y_{1,2})\rangle = 0.$$

Thus $b = Py_{2,2} = \sum_{i,j=1}^{2} e_{i,j} \otimes b_{i,j}$ with $\langle b_{1,2}, v(a)\rangle = \langle b_{2,1}, v(a)\rangle = 0$.

If $\lambda \in C$ is arbitrary with $|\lambda| = 1$, then

$$(3.26) \qquad 2 = \|\lambda y_{1,1} + y_{1,2} + \bar{\lambda} y_{2,2}\|_1 \geq \|\lambda y_{1,1} + y_{1,2} + \bar{\lambda} b\|_1 \geq$$

$$\geq |\langle \lambda y_{1,1} + y_{1,2} + \bar{\lambda} b, v(y_{1,2})\rangle| = 2$$

and thus

$$(3.27) \qquad 2 = \left\|\begin{pmatrix} \bar{\lambda}a+\bar{\lambda}b_{1,1} & a+\bar{\lambda}b_{1,2} \\ a+\bar{\lambda}b_{2,1} & \bar{\lambda}b_{2,2} \end{pmatrix}\right\|_1 = \left\|\begin{pmatrix} 0 & a+\bar{\lambda}b_{1,2} \\ a+\bar{\lambda}b_{2,1} & 0 \end{pmatrix}\right\|_1 = \left\|\begin{pmatrix} 0 & a \\ a & 0 \end{pmatrix}\right\|_1$$

Using Proposition 1.4 we get

$$(3.28) \qquad 0 \leq v(y_{1,2})^*(\lambda_{1,1}+y_{1,2}+\bar{\lambda}b) = \begin{pmatrix} |a|+\bar{\lambda}v(a)^*b_{2,1} & \bar{\lambda}v(a)^*b_{2,2} \\ \lambda|a|+\bar{\lambda}v(a)^*b_{1,1} & \bar{\lambda}v(a)^*b_{1,2}+|a| \end{pmatrix}$$

In particular, this operator is self-adjoint, so:

$$(3.29) \qquad \bar{\lambda}v(a)^*b_{2,2} = \bar{\lambda}|a|+\lambda b_{1,1}^*v(a)$$

and since this holds for arbitrary λ with $|\lambda| = 1$, we get that $b_{1,1} = 0$, and $b_{2,2} = v(a)|a| = a$. So $F(y_{1,1})\cdot Py_{2,2} = y_{2,2} \in R(P)$ by Lemma 3.1.

(ii) Since $E(y_{2,2} + y_{3,3})\, y_{2,3} = y_{2,3}$, we get by Lemma 3.1 that $E(y_{2,2} + y_{3,3})b = b$, where $b = Py_{2,3}$. Also, since $G(y_{2,2})\, y_{2,3} = y_{2,3}$ we get by Lemma 3.1 that $F(y_{2,2})b = 0$. Similarly, $F(y_{3,3})b = 0$, and thus $b = e_{2,3} \otimes b_{2,3} + e_{3,2} \otimes b_{3,2}$. Put $y = \sum_{\substack{1 \leq i \leq j \leq 3 \\ (i,j) \neq (2,3)}} y_{i,j}$, then

$$(3.30) \qquad 3 = \Big\| \sum_{1 \leq i \leq j \leq 3} y_{i,j} \Big\|_1 \geq \|y + b\|_1 \geq |\langle y+b,\ v(y_{1,3}+y_{2,2})\rangle| = 3$$

So equality holds everywhere. By Proposition 1.4 and the equality $\|y+b\|_1 = \|y_{1,3}+y_{2,2}\|_1$ which we just proved, we get

$$(2.31) \qquad 0 \leq v(y_{1,3} + y_{2,2})^*(y + b) = \begin{pmatrix} |a| & v(a)^* b_{3,2} & |a| \\ |a| & |a| & v(a)^* b_{2,3} \\ |a| & |a| & |a| \end{pmatrix}$$

Since this operator is self adjoint, we get $b_{2,3} = b_{3,2} = a$, and thus $b = Py_{2,3} = y_{2,3}$. □

4. RELATIONS BETWEEN ATOMS

In this section we work in C_1, and P will denote a fixed contractive projection in C_1. This section is devoted to the study of the possible relations between atoms of P and/or atoms of contractive subprojections of P, generated by the support projections of elements of $R(P)$ (in the sense of Lemma 3.1).

We begin with a trivial but useful fact whose proof is omitted.

Proposition 4.1: Let $x, y \in B(\ell_2)$. If they satisfy one of the three (exclusive) relations: $E(x)y = y$, $F(x)y = y$ or $G(x)y = y$, then $\ell(x)\,\ell(y) = \ell(y)\,\ell(x)$ and $r(x)\,r(y) = r(y)\,r(x)$. In particular $\{E(x), F(x), G(x), E(y), F(y), G(y)\}$ is a commutative family of projections.

Next, we study the possible relations between an atom x of P and an atom y of $G(x)P$. Note that by Lemma 3.1, $G(x)P$ is a contractive projection in C_1 provided it is not zero. Such a comment will be omitted in the sequel, and we shall use Lemma 3.1 freely to produce such subprojections of P.

Lemma 4.2: Let x be an atom of P and let y be an atom of $G(x)P$. Then either $E(y)x = x$ or $G(y)x = x$.

In the first case there is an $x_0 \in C_1$ and a tensor product representation in which $x = e_{1,1} \otimes x_0, y = (e_{1,2} + e_{2,1})/2 \otimes x_0$ and $z = e_{2,2} \otimes x_0$ is an atom of P.

In the second case there are elements $x_1, x_2 \in C_1$ with $x_1 \perp x_2$ and $\|x_1 + x_2\|_1 = 1$, and there is a tensor product representation in which $x = e_{1,2} \otimes x_1 + e_{2,1} \otimes x_2$, $y = e_{1,3} \otimes x_1 + e_{3,1} \otimes x_2$. Moreover, in this case we have $\|sx + ty\|_1 = (|s|^2 + |t|^2)^{1/2}$ for every scalars s,t aand every normalized element $sx + ty$ is an atom of P; in particular, y is an atom of P.

Proof: The elements $E(y)x$, $F(y)x$ and $G(y)x$ belong to $R(E(x)P)$, and since x is an atom of P we have scalars, λ, μ and ν so that

(4.1) $E(y)x = \lambda x,\ F(y)x = \mu x,\ \ G(y)x = \nu x$

Since $E(y)$, $F(y)$, $G(y)$ are projections whose sum is the identity we have $\lambda,\mu,\nu \in \{0,1\}$ and $\lambda + \mu + \nu = 1$. Let $y_1 = \ell(x)y(1 - r(x))$ and $y_2 = (1 - \ell(x))y\, r(x)$. If $\mu = 1$, we have $x = F(y)x = (\ell(x) - \ell(y_1))x \cdot (r(x) - r(y_2))$ and this clearly implies $\ell(y_1) = 0 = r(y_2)$, which contradicts $y \neq 0$. It follows that $F(y)x = 0$, and thus either $E(y)x = x$ or $G(y)x = x$. In this step we don't use the fact that y is an atom of $G(x)P$.

Assume that $E(y)x = x$, then $\ell(x) = \ell(y_1)$ and $r(x) = r(y_2)$. Put $w = v(y_1)^* v(x)$ and $u = v(y_2)v(x)^*$, then w and u are partial isometries satisfying $r(w) = r(x)$, $\ell(w) = r(y_1)$, $r(u) = \ell(x)$ and $\ell(u) = \ell(y_2)$. Choose any normalized vector $f_1 \in \ell_2$ so that $r(x)f_1 = f_1$, and put $f_2 = wf_1$, $e_1 = v(x)f_1$ and $e_2 = ue_1$, and let us denote $e_{i,j} = (\cdot,f_j)e_i$, $1 \leq i,j \leq 2$. We claim that:

(4.2) $Pe_{1,1} = x$ and $Pe_{1,2} = Pe_{2,1} = y$

Indeed, the first statement is just corolllary 3.4. Since $G(x)e_{1,2} = e_{1,2}$ we get by Lemma 3.1 $F(x)Pe_{1,2} = F(x)P\ F(x)e_{1,2} = 0$, and by Corollary 3.4: $G(x)Pe_{1,2} = y$. So, using Proposition 1.4 we get $E(x)Pe_{1,2} = 0$, and finally $Pe_{1,2} = G(x)Pe_{1,2} = y$. The proof for $e_{2,1}$ is the same. Let $z = Pe_{2,2}$ then by Lemma 3.1

(4.3) $E(y)z = E(y)P\ E(y)e_{2,2} = PE(y)e_{2,2} = z$

(4.4) $G(x)z = G(x)P\ E(y)e_{2,2} = \langle e_{2,2}, v(y)\rangle y = 0$

and

$$(4.5)\quad E(x)z = E(x)P\, e_{2,2} = \langle e_{2,2}, P^* v(x)\rangle = \lambda_0 x$$

where we used the facts that y is an atom of $G(x)P$ and x is an atom of P, respectively. Put $\tilde{z} = F(x)z$, then $z = \lambda_0 x + \tilde{z}$, and if $\lambda \in C$, $|\lambda| = 1$, then by the contractivity of P and (4.2),

$$(4.6)\quad 2 = \left\|\begin{pmatrix} \bar{\lambda} & 1 \\ 1 & \lambda \end{pmatrix}\right\|_1 = \|\bar{\lambda}\, e_{1,1} + e_{1,2} + e_{2,1} + \lambda e_{2,2}\|_1 \geq$$

$$\geq \|(\bar{\lambda} + \lambda_0\lambda)x + 2y + \lambda\tilde{z}\|_1 \geq \|2y\|_1 = 2.$$

Thus, we have equalities everywhere in (4.6), and using Proposition 1.4 we get

$$(4.7)\quad 0 \leq v(y)^*((\bar{\lambda} + \lambda_0\lambda)x + 2y + \lambda z) = \begin{pmatrix} 2|y_2| & \lambda v(y_2)^*\tilde{z} \\ (\bar{\lambda}+\lambda_0\lambda)v(y_1)^* x & 2|y_1| \end{pmatrix}$$

This operator is in particular selfadjoint, so we get

$$(4.8)\quad (\lambda + \overline{\lambda_0\lambda})x^*\, v(y_1) = \lambda v(y_2)^*\tilde{z}$$

In particular, $|\lambda+\overline{\lambda_0\lambda}| = \|(\lambda+\overline{\lambda_0\lambda})x^* v(y_1)\|_1 \leq 1$, and since λ is arbitrary we get $\lambda_0 = 0$ and thus

$$(4.9)\quad Pe_{2,2} = z = \tilde{z} = v(y_2)x^* v(y_1) = u\, x\, w^*.$$

If $b \in R(E(z)P) \neq 0$, then $G(b)y = ty$ for some scalar t (since y is an atom of $G(x)P$), so necessarily $E(b) = E(z)$.

We get therefore that z is an atom of P. Finally, using Proposition 1.2 we get

$$(4.10)\quad 4 = \left\|\begin{pmatrix} 1 & 1 \\ 1 & 1 \end{pmatrix}\right\|_1^2 = \left\|\sum_{i,j=1}^{2} e_{i,j}\right\|_1^2 \geq \left\|P\sum_{i,j=1}^{2} e_{i,j}\right\|_1^2 = \left\|\begin{pmatrix} x & 2y_1 \\ 2y_2 & z \end{pmatrix}\right\|_1^2 \geq$$

$$\geq \|x\|_1^2 + 4\|y_1\|_1^2 + 4\|y_2\|_1^2 + \|z\|_1^2 = 2 + 4(\|y_1\|_1^2 + (1 - \|y_1\|_1)^2) =$$

$$= 4 + 8(\|y_1\|_1 - 1/2)^2$$

Thus, $\|y_1\|_1 = \|y_2\|_1 = 1/2$ and we have equality everywhere in (4.10). By the equality case of Proposition 1.2 we get

$$(4.11) \quad y_1 = xw^*/2 \quad ; \quad y_2 = ux/2$$

Now from (4.9) and (4.11) it is clear how to choose $x_0 \in C_1$ and a tensor product representation so that

$$(4.12) \quad x = e_{1,1} \otimes x_0, \quad y = (e_{1,2} + e_{2,1})/2 \otimes x_0, \quad z = e_{2,2} \otimes x_0$$

and this completes the proof in the case $E(y)x = x$.

Assume that $G(y)x = x$, and let $x_1 = \ell(y_1)x$, $x_2 = xr(y_2)$. Clearly, $x = x_1 + x_2$ and $x_1 \perp x_2$. We assume for definiteness that $x_1 \neq 0 \neq x_2$, the proof in the case when one of the x_j vanish is the same. Let $w = v(y_1)^* v(x_1)$, $u = v(y_2) v(x_2)^*$; then w and u are partial isometries satisfying $r(w) = r(x_1)$, $\ell(w) = r(y_1)$, $r(u) = \ell(x_2)$ and $\ell(u) = \ell(y_2)$. Let f_1 and f_2 be normalized vectors in ℓ_2 so that $r(x_2)f_1 = f_1$, $r(x_1)f_2 = f_2$ and let us define $f_3 = wf_2$, $e_1 = v(x_1)f_2$, $e_2 = v(x_2)f_1$, $e_3 = ue_2$. Let $e_{i,j} = (\cdot, f_j)e_i$, $1 \leq i,j \leq 3$. Then, as in the proof of the first case:

$$(4.13) \quad P\, e_{1,2} = P\, e_{2,1} = x; \; P\, e_{1,3} = P\, e_{3,1} = y$$

Using Proposition 1.2 we get

$$(4.14) \quad 8 = \left\| \begin{pmatrix} 0 & 1 & 1 \\ 1 & 0 & 0 \\ 1 & 0 & 0 \end{pmatrix} \right\|_1^2 = \|e_{1,2} + e_{1,3} + e_{2,1} + e_{3,1}\|_1^2 \geq$$

$$\geq \|P(e_{1,2} + e_{1,3} + e_{2,1} + e_{3,1})\|_1^2 = \|2x + 2y\|_1^2 = 4\left\|\begin{pmatrix} 0 & x_1 & y_1 \\ x_2 & 0 & 0 \\ y_2 & 0 & 0 \end{pmatrix}\right\|_1^2 \geq$$

$$\geq 4((\|x_1\|_1^2 + \|y_1\|_1^2)^{1/2} + (\|x_2\|_1^2 + \|y_2\|_1^2)^{1/2})$$

Since $\|x_1\|_1 + \|x_2\|_1 = 1 = \|y_1\|_1 + \|y_2\|_1$ one can easily deduce from (4.14) that $\|x_1\|_1 = \|y_1\|_1$, $\|x_2\|_1 = \|y_2\|_1$, and thus we have in (4.14) equalities everywhere. Using the equality case in Proposition 1.2 we get

(4.15) $y_1 = x_1 w^*$; $y_2 = ux_2$

If s,t are scalars then (4.15) easily implied that

(4.16) $\|sx + ty\|_1 = (|s|^2 + |t|^2)^{1/2}$

In order to show that y is an atom of P it is clearly enough to show that $Q = F(x)\,E(y)P = 0$, since y is known to be an atom of $G(x)P$. If $Q \neq 0$, let a be any element of $R(Q)$. Since y is an atom of $G(x)P$ we must have $G(a)y = y$, so $E(a) = F(x)\,E(y)$, and thus a is an atom of P and y is an atom of $G(a)P$. By the first case of the present lemma, we get $F(a)E(y)P \neq 0$ and this contradicts the atomocity of x, since $F(a)E(y) < E(x)$. So, indeed, $Q = 0$ and y is an atom of P.

Let now $z = sx + ty$, $\|z\|_1 = 1$ with $s \neq 0 \neq t$. If $0 \neq a \in R(E(z)P)$, then by the atomocity of x and y we have $E(x)a = \sigma x$ and $E(y)a = \tau y$, and so

(4.17) $$a = \begin{pmatrix} 0 & \sigma x_1 & \tau y_1 \\ \sigma x_2 & 0 & b_1 \\ \tau y_2 & b_2 & 0 \end{pmatrix}, \quad \text{where } b = b_1 + b_2 \in R(P).$$

If $b \neq 0$, we must have $G(b)y = y$ and $G(b)x = x$, so

$r(b) = r(x_1) + r(y_1)$ and $\ell(b) = \ell(x_2) + \ell(y_2)$. Now $r(a) \leq r(z)$, so $\ker(sx_1 + ty_1) \subseteq (\ker(b) \cap \ker(\sigma x_1 + \tau y_1))$. Let $\xi = tf_2 - sf_3$. Then $0 \neq \xi \in \ker(sx_1 + ty_1)$ and $b\xi = 0$ contradicts $b \neq 0$, so $b = 0$ and $a = \sigma x + \tau y$. Also, $a\xi = 0$ implies using (4.15) that for some $\lambda \in C$, $\sigma/s = \tau/t = \lambda$, and so $a = \lambda z$ and z is an atom of P.

Finally, (4.15) is clearly equivalent to the existence of a tensor product representation in which $x = e_{1,2} \otimes x_1 + e_{2,1} \otimes x_2$ and $y = e_{1,3} \otimes x_1 + e_{3,1} \otimes x_2$. □

Remark 4.3: Note that the two possibilities in Lemma 4.2 are exclusive in the following sense: if x is an atom of P and $y, \tilde{y}$ are atoms of $G(x)P$, then either $G(y)x = G(\tilde{y})x = x$ or $E(y)x = E(\tilde{y})x = x$. Indeed, if for example $G(y)x = x = E(\tilde{y})x$, then $E(\tilde{y})y = \lambda\tilde{y} = y_1\, r(\tilde{y}_1) + \ell(\tilde{y}_2)y_2$, where $y = y_1 + y_2$ and $\tilde{y} = \tilde{y}_1 + \tilde{y}_2$ are the decompositions appearing in the proof of Lemma 4.2, and $\lambda \in C$, and thus $\lambda\tilde{y}_1 = y_1 \cdot r(\tilde{y}_1)$ and $\lambda\tilde{y}_2 = \ell(\tilde{y}_2)y_2$. If $\lambda \neq 0$ then $E(\tilde{y})x \neq x$; if $\lambda = 0$ then $0 \neq G(y)\tilde{y} \in R(E(\tilde{y})G(x)P)$, so $G(y)\tilde{y} = \tilde{y}$ and again $E(\tilde{y})x \neq x$. This contradiction, establishes our claim.

Note also that in the first case of Lemma 4.2, y is not an atom of P.

In order to continue our analysis we need one more concept:

Definition 4.4: A system $(x_1, x_2; \tilde{x}_1, \tilde{x}_2)$ of normalized elements of C_1 is said to satisfy the matrix condition (condition "M", for short), if there exists elements $a, b \in C_1$ with $a \perp b$, and there exists partial isometries u, τ, w and σ with

(4.18) $r(u) = \ell(a)$, $r(\tau) = \ell(b)$; $r(w) = r(a)$, $r(\sigma) = r(b)$;

(4.19) $\{r(u), \ell(u), r(\tau), \ell(\tau)\}$ are pairwise orthogonal;

(4.20) $\{r(w), \ell(w), r(\sigma), \ell(\sigma)\}$ are pairwise orthogonal;

and so that

(4.21) $x_1 = a + b;\ x_2 = aw^* + \tau b$

(4.22) $\tilde{x}_1 = uaw^* + \tau b\sigma^*\ ;\quad \tilde{x}_2 = -\ (ua + b\sigma^*)$

<u>Remark 4.5:</u> It is clear that the system $(x_1,x_2;\tilde{x}_1,\tilde{x}_2)$ satisfies condition M if and only if there exist a matrix representation in which

(4.23) $x_1 = \begin{bmatrix} a & 0 & 0 & 0 \\ 0 & 0 & 0 & 0 \\ 0 & 0 & 0 & 0 \\ 0 & 0 & 0 & b \end{bmatrix}$ $\qquad x_2 = \begin{bmatrix} 0 & a & 0 & 0 \\ 0 & 0 & 0 & 0 \\ 0 & 0 & 0 & 0 \\ 0 & 0 & b & 0 \end{bmatrix}$

(4.24) $\tilde{x}_1 = \begin{bmatrix} 0 & 0 & 0 & 0 \\ 0 & a & 0 & 0 \\ 0 & 0 & 0 & 0 \\ 0 & 0 & 0 & b \end{bmatrix}$ $\qquad \tilde{x}_2 = \begin{bmatrix} 0 & 0 & 0 & 0 \\ -a & 0 & 0 & 0 \\ 0 & 0 & 0 & -b \\ 0 & 0 & 0 & 0 \end{bmatrix}$

where $\|a\|_1 + \|b\|_1 = 1$. In tensor product notation this can be written as

(4.25) $x_1 = e_{1,1} \otimes a + e_{1,1} \otimes b;\quad x_2 = e_{1,2} \otimes a + e_{2,1} \otimes b$

(4.26) $\tilde{x}_1 = e_{2,2} \otimes a + e_{2,2} \otimes b;\quad \tilde{x}_2 = -(e_{2,1} \otimes a + e_{1,2} \otimes b)$

If the system $(x_1,x_2;\tilde{x}_1,\tilde{x}_2)$ satisfies M, then (trivially) the systems $(\tilde{x}_1,\tilde{x}_2;x_1,x_2)$ and $(x_2,x_1;\tilde{x}_2,\tilde{x}_1)$ satisfy M. Also, if $(x_1,x_2;\tilde{x}_1,\tilde{x}_2)$ satisfies M and if $t_1,t_2;\tilde{t}_1,\tilde{t}_2$ are scalars of absolute value 1, then the system

(4.27) $(t_1x_1,t_2x_2;\ \tilde{t}_1\tilde{x}_1,\ \tilde{t}_2\tilde{x}_2)$

satisfied M provided $t_1\tilde{t}_1 = t_2\tilde{t}_2$.

In section 6 we shall characterize systems satisfying M in terms of isometries.

Let us turn back to the study of the atoms of the contractive projection P in C_1.

Proposition 4.6: Let x,y be atoms of P with x G y. Then $F(x)G(y)P \neq 0$ if and only if $F(y)G(x)P \neq 0$. If $F(x)G(y)P \neq 0$ then there exist atoms $\tilde{x},\tilde{y}$ of P so that $(x,y;\tilde{x},\tilde{y})$ satisfies condition M.

Proof: Put $w = v(y)^* v(x)$ and $\tau = v(y)v(x)^*$. Then for some $a,b \in C_1$ with $a \perp b$ and $\|a+b\|_1 = 1$ we have

$$(4.28) \quad x = a + b,\ y = aw^* + \tau b\ .$$

Assume that $F(y)G(x)P \neq 0$, and let $\tilde{y}$ be an atom of $F(y)G(x)P$. As in the beginning of the proof of Lemma 4.2, we get $G(\tilde{y})x = x$, so $x\ G\ \tilde{y}$. Since $E(\tilde{y})G(x) = E(\tilde{y})F(y)G(x)$, we get that $\tilde{y}$ is an atom of $G(x)P$, and thus by Lemma 4.2 $\tilde{y}$ is an atom of P. Put $u = -\ v(\tilde{y})v(x)^*$, $\sigma = -v(\tilde{y})^* v(x)$. Then, clearly (4.18), (4.19) and (4.20) are satisfied, and also:

$$(4.24) \quad \tilde{y} = -\ (ua + b\sigma^*)$$

Using part (i) of Proposition 3.6 we get that

$$(4.30) \quad \tilde{x} = uaw^* + \tau b\sigma^* \in \mathbf{R}(P),$$

and thus $F(x)G(y)P \neq 0$. The preceeding analysis concerning $\tilde{y}$ yields by symmetry that if z is any atom of $F(x)G(y)P$, then z is proportional to $\tilde{x}$. Thus, $\tilde{x}$ is an atom of $F(x)G(y)P$, and thus - an atom of P. Obviously, $(x,y;\tilde{x},\tilde{y})$ satisfies the matrix condition. This completes the proof of the proposition in view of the symmetry between x and y.

□

We turn now to the study of atoms of $G(y)G(x)P$. We consider first the alternative "$F(x)G(y)P = 0$" of Proposition 4.6. We use in the following Lemma the matrices $x_{n,k,m}$ of section 2b.

Lemma 4.7: Let x,y be atoms of P with $x \, G \, y$, and assume that $F(x)G(y)P = 0 = F(y)G(x)P$. Let z be an atom of $G(y)G(x)P$. Then z is an atom of P which satisfies $(t_1x + t_2y) G \, z$ for any scalars t_1, t_2 with $|t_1| + |t_2| \neq 0$. So for every scalar t_1, t_2 and t_3:

$$(4.31) \quad \|t_1x + t_2y + t_3z\|_1 = (|t_1|^2 + |t_2|^2 + |t_3|^2)^{1/2}.$$

Also,

$$(4.32) \quad F(x)G(z)P = F(z)G(x)P = F(y)G(z)P = F(z)G(y)P = 0.$$

Moreover, there exist $a_1, a_2, a_3 \in C_1$ with $\|a_1\|_1 + \|a_2\|_1 + \|a_3\|_1 = 1$ and there exists a tensor product representation in which

$$(4.33) \quad x = h_{3,1}(a_1, a_2, a_3) = \sum_{m=1}^{3} x_{3,1,m} \otimes a_m$$

$$(4.34) \quad y = h_{3,2}(a_1, a_2, a_3) = \sum_{m=1}^{3} x_{3,2,m} \otimes a_m$$

$$(4.35) \quad z = h_{3,3}(a_1, a_2, a_3) = \sum_{m=1}^{3} x_{3,3,m} \otimes a_m$$

Proof: Let $x' = \ell(y)x$, $x'' = x\,r(y)$, $y' = \ell(x)y$, $y'' = yr(x)$, $w = v(y')^*v(x')$ and $u = v(y'')v(x'')^*$. By Lemma 4.2 we have

$$(4.36) \quad y' = x'w^*, \; y'' = ux''$$

Put also $z_1 = \ell(y')\ell(x')z$, $z_2 = \ell(x')\, z\, r(y'')$, $z_3 = \ell(y'')z\, r(x')$ and $z_4 = z\, r(x'')r(y'')$. Then $z = z_1 + z_2 + z_3 + z_4$, and our situation can be described matrically as

(4.37)

0	x'	y'	z_1	
x''	0	z_2	0	0
y''	z_3	0	0	0
z_4	0	0		
	0	0		

where a "0" in a square means that no element of $R(P)$ has a non zero component in this square. We assume that $x' \neq 0 \neq x''$, since if one of x' and x'' vanish the proof is almost the same, but easier.

Now x is an atom of $\tilde{P} = G(y)P$ and z is an atom of $G(x)\tilde{P} = G(x)G(y)P$. If $E(z)x = x$, we get by Lemma 4.2 that $E(x)F(x)\tilde{P} \neq 0$, but this is a contradiction since $E(z)F(x)\tilde{P} = E(z)F(x)G(y)P = 0$. So, by Lemma 4.2, $x \; G \; z$, and z is an atom of $\tilde{P} = G(y)P$. Also, by symmetry, $y \; G \; z$, and thus by Lemma 4.2 z is an atom of P.

Thus, we have decompositions $x = \sum_{j=1}^{4} x_j$, $y = \sum_{j=1}^{4} y_j$ where:

(4.38) $\quad x_1 = \ell(z_1)x, \; x_2 = x \; r(z_3), \; x_3 = \ell(z_2)x, \; x_4 = x \; r(z_4);$

(4.39) $\quad y_1 = \ell(z_1)y, \; y_2 = y \; r(z_2), \; y_3 = \ell(z_3)y, \; y_4 = y \; r(z_4);$

(4.40) $\quad |x_1^*| = |y_1^*| = |z_1^*| \; ; \; |x_4| = |y_4| = |z_4| \; ;$

(4.41) $\quad |x_2^*| = |y_2^*| \; , \; |x_3^*| = |z_2^*| \; , \; |y_3^*| = |z_2^*| \; ;$

(4.42) $\quad |x_2| = |z_3|, \; |x_3| = |y_3|, \; |y_2| = |z_2|.$

The following matrix illustrates the mutual relations between the components of x, y and z:

(4.43)

	0	0			0	0	
x_4		0	0	0		0	0
y_4	0		0	0	0		0
z_4	0	0		0	0	0	
0	y_2	x_2	0		0	0	
0	z_2	0	x_3	0		0	0
0	0	z_3	y_3	0	0		0
0	0	0	0	z_1	y_1	x_1	

where the zeros have the meaning as before.

To prove (4.32), it is enough by symmetry to prove for example that $F(z)G(x)P = 0$. Now y and z are atoms of $G(x)P$, and $G(z)F(y)(G(x)P) = 0$. Hence, by Proposition 4.6, $G(y)F(z)(G(x)P) = 0$. So

$$(4.44) \quad \begin{aligned} F(z)G(x)P &= E(y)F(z)G(x)P + F(y)F(z)G(x)P + G(y)F(z)G(x)P \\ &= E(y)F(z)G(x)P + 0 + 0 = 0 \end{aligned}$$

where in the last step we use $E(y)F(z)P = 0$, which is an easy consequence of $y \, G \, z$ and the atomicity of y.

Let $b = t_1 x + t_2 y$, where $|t_1|^2 + |t_2|^2 = 1$. By Lemma 4.2, b is an atom of P. Also, $x \, G \, z$ and $y \, G \, z$ imply that $G(z)b = b$. Now, $E(b)z = z$ is impossible, since b is an atom of P, so necessarily $G(b)z = z$ and thus $b \, G \, z$. This fact and Lemma 4.2 easily imply formula (4.31).

Let us prove (4.33), (4.34) and (4.35). Put

(4.45) $\tau_1 = v(y_4)v(x_4)^*$, $\tau_2 = v(z_4)v(x_4)^*$, $a_1 = x_4$

(4.46) $\sigma_1 = -v(y_1)^*v(x_1)$, $\sigma_2 = v(z_1)^*v(x_1)$, $a_3 = x_1$

Then (4.40) implies that

(4.47) $y_4 = \tau_1 a_1$, $z_4 = \tau_2 a_1$; $y_3 = -a_3\sigma_1^*$, $z_3 = a_3\sigma_2^*$.

In order to complete the proof we need the following proposition.

Proposition 4.8: Let $b,c,d \in C_1$ satisfy

(4.48) $b \, G \, c$, $b \, G \, d$, $c \, G \, d$,

(4.49) $\ell(b)\,\ell(c)\,d = 0$, $d\,r(b)\,r(c) = 0$.

Let $b = b_1 + b_2$, $c = c_1 + c_2$, and $d = d_1 + d_2$ be the decompositions induced by (4.48) and (4.49), that is

(4.50) $b_1 = \ell(c)b = br(d)$, $b_2 = \ell(d)b = br(c)$,

(4.51) $c_1 = \ell(b)c = c\,r(d)$, $c_2 = \ell(d)c = cr(b)$,

(4.52) $d_1 = \ell(b)d = dr(c)$, $d_2 = \ell(c)d = dr(b)$.

Further, assume that

(4.53) $|b_1^*| = |c_1^*|$, $|b_2^*| = |d_1^*|$, $|c_2^*| = |d_2^*|$

(4.54) $|b_1| = |d_2|$, $|b_2| = |c_2|$, $|c_1| = |d_1|$

and that

(4.55) $E(b+c)d = 0$

Then, there exists $a \in C_1$ and partial isometries w_1, w_2, u_1, u_2 so that

(4.56) $r(w_1) = r(w_2) = r(a); \quad r(u_1) = r(u_2) = \ell(a)$

(4.57) $\ell(w_1) = r(b_1), \ \ell(w_2) = r(c_1); \ \ell(u_1) = \ell(b_2), \ \ell(u_2) = \ell(c_2)$

and so that

(4.58) $b_1 = aw_1^*, \quad c_1 = -aw_2^*, \quad d_1 = -u_1 a w_1^*$

(4.59) $b_2 = u_1 a, \quad c_2 = u_2 a, \quad d_2 = -u_2 a w_1^*$

We first complete the proof of Lemma 4.7 using Proposition 4.8. Let $b_j = x_{j-1}$, $c_j = y_{j-1}$ and $d_j = z_{j-1}$, $j = 1,2$, Then $b = b_1 + b_2$, $c = c_1 + c_2$ and $d = d_1 + d_2$ satisfy all the relations (4.48) - (4.55). Indeed, (4.48) - (4.54) follow from the proof of the lemma and the definition of b, c, and d; and (4.55) follows from $G(x+y)z = z$. Put $a_2 = a$, where a is that of Proposition 4.8. Then the relations (4.58) and (4.59) can be written in a suitable tensor product representation as

(4.60) $x_2 + x_3 = x_{3,1,2} \otimes a_2, \ y_2 + y_3 = x_{3,2,2} \otimes a_2, \ z_2 + z_3 = x_{3,3,2} \otimes a_2$.

Thus, (4.33), (4.34) and (4.35) follow from (4.47) and (4.60).

Proof of Proposition 4.8: Let us define

(4.61) $w = -v(c_1)^* v(b_1), \ u = v(c_2)v(b_2)^*$

Then, from (4.53) and (4.54) we get

(4.62) $v(b_1 + c_1) = (v(b_1) - v(b_1)w^*)/\sqrt{2} \ ; \quad v(b_2 + c_2) = (v(b_2) + uv(b_2))/\sqrt{2}$

Now (4.55) is equivalent in our case to

(4.63) $\ell(b_2 + c_2) \cdot d \cdot r(b_1 + c_1) = 0.$

So

(4.64) $0 = v(b_2 + c_2)^* \ell(b_2 + c_2)\ d\ r(b_1 + c_1)\ v(b_1 + c_1)^*$

$= v(b_2 + c_2)^* \cdot d \cdot v(b_1 + c_1)^*$

$= (v(b_2)^* + v(b_2)^* u^*) \cdot (d_1 + d_2) \cdot (v(b_1)^* - wv(b_1)^*)/2$

$= (-v(b_2)^* d_1 w\ v(b_1)^* + v(b_2)^* u^* d_2\ v(b_1)^*)/2$

Hence

(4.65) $d_2 = ud_1 w$

Put

(4.66) $w_1 = -v(d_2)^* v(c_2),\ w_2 = ww_1$

(4.67) $u_1 = v(d_1)v(c_1)^*,\ uu_2 = uu_1$

(4.68) $a = u_1^* b_2$

The relations (4.59) follow directly from the definition of a, the definitions of the partial isometries and from (4.53) and (4.54). Now, from (4.65) we get

(4.69) $d_1 = u^* d_2 w^* = -u^* u_2 a w^*{}_1 w^* = -u_1\ a\ w_2^*$.

From this, the remainder relations in (4.58) followbyby the definition of u_1, w and by (4.53) and (4.54).

This completes the proof of Proposition 4.8, and thus the proof of Lemma 4.7. □

Let us continue our analysis with the atoms x,y of P so that $x\ G\ y$, in the alternative "$F(x)G(y)P \neq 0$" of Proposition 4.6. By Proposition 4.6 there exist atoms $\tilde{x},\tilde{y}$ of P so that the system $(x,y;\tilde{x},\tilde{y})$ satisfies M. Put

(4.70) $Q = G(\tilde{x})\ G(y)\ G(x)P$

Clearly, $G(\tilde{y})Q = Q$. If $Q \neq 0$, then the structure of its atoms depend on whether $P_1 = G(x+\tilde{x})P = 0$ or not. Let us consider first the case $P_1 \neq 0$.

Now, P_1 is the sum of four subprojections.

$$(4.71) \quad P_1 = G(y)G(x)P_1 + G(y)G(\tilde{x})P_1 + G(\tilde{y})G(\tilde{x})P_1 + G(\tilde{y})G(x)P_1$$

Thus, the study of the atoms of P_1 reduces to the study of the atoms of these subprojections. The following diagram describes matricially these subprojections, as well as the projection Q and the elements $x,y,\tilde{x}$ and $\tilde{y}$;

(4.72)

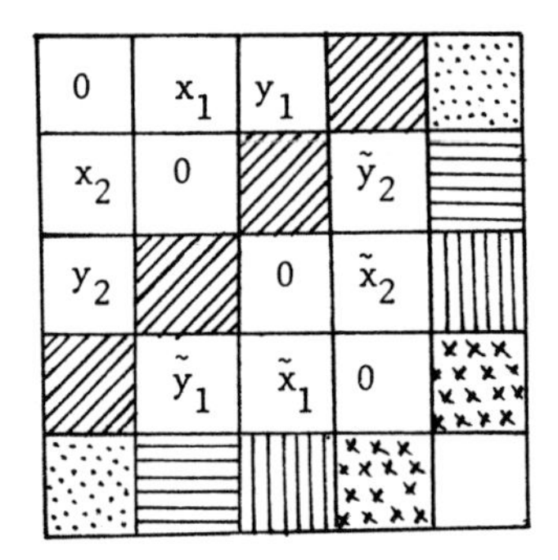

Here $x_i, y_i, \tilde{x}_i$ and $\tilde{y}_i$ are the components of $x,y,\tilde{x}$ and $\tilde{y}$, and the shaded areas , , , and correspond to the projections Q, $G(y)\,G(x)\,P_1$, $G(\tilde{y})G(x)P_1$, $G(y)G(\tilde{x})P_1$ and $G(\tilde{y})G(\tilde{x})P_1$ respectively.

Now, an application of Proposition 3.6 gives us that:

$$(4.72) \quad G(y)G(x)P_1 \neq 0 \text{ if and only if } G(\tilde{y})G(\tilde{x})P_1 \neq 0$$

and

$$(4.73) \quad G(\tilde{y})G(x)P_1 \neq 0 \text{ if and only if } G(y)G(\tilde{x})P_1 \neq 0.$$

In view of the symmetry between x and $\tilde{x}$ (and between y and $\tilde{y}$) it is enough to study only one case, for example $G(y)G(x)P_1 \neq 0$.

<u>Proposition 4.9:</u> Let $x,y,\tilde{x},\tilde{y}$ be atoms of P so that $(x,y;\tilde{x},\tilde{y})$ satisfies M, let $P_1 = G(x+\tilde{x})P$ and let z be any atom of $G(y)G(x)P_1$. Let

(4.75) $Q_1 = G(z)Q, \quad Q_2 = F(z)Q$

where Q is defined by (4.70). Then $Q_1 \neq 0$ if and only if $Q_2 \neq 0$. Moreover, in the case $Q_1 \neq 0 \neq Q_2$ there exist a tensor product representation in which for some $a \in C_1$ with $\|a\|_1 = 1/2$ all the elements

(4.76) $y_{i,j} = a_{i,j} \otimes a, \quad 1 \leq i < j \leq 5$

are atoms of P, where $a_{i,j}$ are the elementary anti-symmetric matrices, and

(4.77) $x = y_{1,2}\ , \ y = y_{1,3}\ , \ \tilde{x} = y_{3,4}, \ \tilde{y} = -y_{2,4}, \ z = y_{1,5}.$

Proof: We clearly have $G(z)x = x$, $G(z)y = y$ and $E(z)E(x)P = E(z)E(y)P = 0$. So $E(z)G(y)G(x)P = E(z)G(y)G(x)P_1$, and thus z is an atom of $G(y)G(x)P$. Now y is an atom of $\tilde{P} = G(x)P$ (being an atom of P), z is an atom of $G(y)\tilde{P} = G(y)G(x)P$ with zGy, so by Lemma 4.2 z is an atom of $G(x)P = \tilde{P}$. Using again Lemma 4.2 with P, x and z, we get from zGx that z is an atom of P.

Assume that $Q_2 \neq 0$ and let b be an atom of Q_2. Clearly bGx, bGy and $E(b)Q_2 = E(b)G(y)G(x)P$. So, b is an atom of $G(y)G(x)P$, and exactly as before we get from this that b is an atom of P. The relations bGx and bGy implies also that up to multiplication by scalar of absolute value one, b is the only atom of Q_2. Thus we must have $E(x+y)b = 0$. Similarly, $bG\tilde{y}$, $bG\tilde{x}$ and $E(b+\tilde{y})\tilde{x} = 0$.

Using the relations bGx and bGy, Lemma 4.2, the fact that $E(x+y)b = 0$ we get by Proposition 4.8 the existence of some $a \in C_1$ with $\|a\|_1 = 1/2$, and the existence of partial isometries w_1, w_2, u_1, u_2 so that:

(4.78) $r(u_1) = r(u_2) = \ell(a)\ ; \quad r(w_1) = r(w_2) = r(a);$

(4.79) $\{r(a), \ell(w_1), \ell(w_2)\}$ are pairwise orthogonal;

(4.80) $\{\ell(a), \ell(u_1), \ell(u_2)\}$ are pairwise orthogonal ;

(4.81) $x = aw_1^* - u_1a$; $y = aw_2^* - u_2a^*$

(4.82) $b = u_1aw_2^* - u_2aw_1^*$.

Put

(4.83) $w = -v(\tilde{y})^* v(b)$, $u = -v(\tilde{y})v(b)^*$, $w_3 = ww_2$, $u_3 = uu_2$.

Then by Proposition 4.8 applied to b, $\tilde{y}$ and $\tilde{x}$, and by the fact that $(x,y;\tilde{x},\tilde{y})$ satisfies M, we get

(4.84) $\tilde{y} = -u_1aw_3^* + u_3aw_1^*$; $\tilde{x} = u_2aw_3^* - u_3aw_2^*$

Let also

(4.85) $w_4 = v(z)^* v(x)w_1$, $u_4 = v(z)v(x)^* u_1$

Then

(4.86) $z = aw_4^* - u_4a$

Put $y_{1,2} = x$, $y_{1,3} = y$, $y_{1,5} = z$, $y_{2,3} = b$, $y_{2,4} = -\tilde{y}$ and $y_{3,4} = \tilde{x}$. Applying Proposition 3.6 to the triplets, $\{y_{1,3}, y_{2,3}, y_{2,4}\}$, $\{y_{1,3}, y_{1,5}, y_{2,3}\}$, $\{y_{1,2}, y_{1,5}, y_{2,3}\}$ and $\{y_{1,2}, y_{1,5}, y_{2,4}\}$ we get in $R(P)$ the following elements respectively.

(4.87) $y_{1,4} = aw_3^* - u_3a$

(4.88) $y_{2,5} = u_1aw_4^* - u_4aw_1^*$

(4.89) $y_{3,5} = u_2aw_4^* - u_4aw_2^*$

and

(4.90) $y_{4,5} = u_3aw_4^* - u_4aw_3^*$

Also, by the proof of Proposition 4.4, $y_{1,4}$, $y_{2,5}$, $y_{3,5}$ and $y_{4,5}$ are atoms of P, and $y_{4,5}$ is an atom of Q_1, so $Q_1 \neq 0$. Clearly, the relations (4.81), (4.82), (4.84), (4.86), (4.87), (4.88) and (4.89) are equivalent to the existence of a tensor product representation in which (4.76) holds.

Conversely, assume that $Q_1 \neq 0$, and let c be an atom of Q_1. By the atomocity of y and $\tilde{y}$ we get cGy and $cG\tilde{y}$. Also $E(c)Q_1 = E(c)P$, since $E(c)E(\tilde{x}) = E(c)E(x) = 0$. So c is an atom of P, and thus an atom of $G(\tilde{y})P$ and of $G(y)P$. Using the conclusion of Lemma 4.2 to the pairs $\{c,y\}$ and $\{c,\tilde{y}\}$ in Proposition 3.6, we get that $F(c)G(y)G(\tilde{y})P = Q_2 \neq 0$. □

Next, let us continue our study of the atoms of Q under the assumption $P_1 = 0$.

Lemma 4.10: Let x,y, $\tilde{x},\tilde{y}$ be atoms of P so that the system $(x,y;\tilde{x},\tilde{y})$ satisfies M, let $Q = G(\tilde{x})G(y)G(x)P$ and assume that $P_1 = G(x+\tilde{x})P = 0$. If z is any atom of Q then either $E(z) = E(x+\tilde{x})$ and

(4.91) $E(z)x = x$, $E(z)y = y$, $E(z)\tilde{x} = \tilde{x}$, $E(z)\tilde{y} = \tilde{y}$;

or $E(z) < E(x+\tilde{x})$ and

(4.92) zGx, zGy, $zG\tilde{x}$, $zG\tilde{y}$.

In the first case z can be chosen so that there exists a tensor product representation and an element $a_o \in C_1$ with $\|a_o\|_1 = 1/2$ and:

(4.93) $x = x_{3,1} \otimes a_o$; $y = x_{3,2} \otimes a_o$

(4.94) $\tilde{x} = \tilde{x}_{3,1} \otimes a_o$; $\tilde{y} = \tilde{x}_{3,2} \otimes a_o$

(4.95) $z = (x_{3,3} + \tilde{x}_{3,3})/2 \otimes a_o$

where the matrices $x_{3,k}$ and $\tilde{x}_{3,k}$, il $\leq k \leq 3$, are defined by (2.38) and (2.39) respectively.

In the second case there exists besides z another atom of Q denoted by $\tilde{z}$, so that $z \perp \tilde{z}$, z and $\tilde{z}$ are atoms of P and (4.92) holds with $\tilde{z}$ instead of z. Moreover, z and $\tilde{z}$ can be chosen so that there exist $a,b \in C_1$ with $\|a\|_1 + \|b\|_1 = 1/2$ and there exists a tensor product representation in which

$$(4.96) \quad x = x_{3,1} \otimes a + (\tilde{x}_{3,1})^T \otimes b; \quad \tilde{x} = \tilde{x}_{3,1} \otimes a + x_{3,1}^T \otimes b \ ;$$

$$(4.97) \quad y = x_{3,2} \otimes a + (\tilde{x}_{3,2})^T \otimes b; \quad \tilde{y} = \tilde{x}_{3,2} \otimes a + x_{3,2}^T \otimes b \ ;$$

$$(4.98) \quad z = x_{3,3} \otimes a + (\tilde{x}_{3,3})^T \otimes b; \quad \tilde{z} = \tilde{x}_{3,3} \otimes a + x_{3,3}^T \otimes b \ ;$$

where the $x_{3,k}$ and $\tilde{x}_{3,k}$, $1 \leq k \leq 3$, are as above and "T" means transpose. Thus each of the following systems satisfies M: $(x,z;\tilde{x},\tilde{z})$ and $(y,z;\tilde{y},\tilde{z})$.

Proof: y is an atom of $G(\tilde{x})G(x)P = \tilde{P}$ and z is an atom of $G(y)\tilde{P} = Q$. So, by Lemma 4.2, either $E(z)y = y$ or $zG\tilde{y}$. In the first case we cannot have $zG\tilde{y}$ and thus, by the symmetry between y and $\tilde{y}$, we get also $E(z)\tilde{y} = \tilde{y}$. This implies $E(z) = E(x+\tilde{x}) = E(y+\tilde{y})$, $E(z)x = x$ and $E(z)\tilde{x} = \tilde{x}$. In the second case, namely zGy, we cannot have any of the relations $E(z)x = x$, $E(z)\tilde{x} = \tilde{x}$, $E(z)y = y$, since any one of them imply by the proof in the first case (and by the symmetry between $x,y,\tilde{x}$ and $\tilde{y}$) that $E(z)y = y$, a contradiction. So (4.92) holds.

Assume now the first case. Put

$$(4.99) \quad x_1 = \ell(y)x, \quad x_2 = \ell(\tilde{y})x; \quad y_1 = \ell(x)y, \quad y_2 = \ell(\tilde{x})y;$$

$$(4.100) \quad \tilde{x}_1 = \ell(y)\tilde{x}, \quad \tilde{x}_2 = \ell(\tilde{y})\tilde{x}; \quad \tilde{y}_1 = \ell(x)\tilde{y}, \quad \tilde{y}_2 = \ell(\tilde{x})\tilde{y};$$

(4.101) $$z_1 = \ell(y)\,\ell(x)z,\ z_2 = \ell(\tilde{y})\,\ell(x)z,\quad z_3 = \ell(y)\,\ell(\tilde{x})z,\quad z_4 = \ell(\tilde{y})\ell(\tilde{x})z.$$

clearly, $x = x_1 + x_2$, $y = y_1 + y_2$, $\tilde{x} = \tilde{x}_1 + \tilde{x}_2$, $\tilde{y} = \tilde{y}_1 + \tilde{y}_2$ and $z = z_1 + z_2 + z_3 + z_4$. We can choose a matrix representation which shows the relations between the components of $x,y,\tilde{x},\tilde{y}$ and z:

(4.102)

x_1	$\tilde{y}_1$	z_2	0
y_1	$\tilde{x}_1$	0	z_2
z_4	0	$\tilde{x}_2$	$\tilde{y}_2$
0	z_1	y_2	x_2

Here, the support of the component described in (4.102) are "exact" (for example, $r(z_4) = r(y_1)$, etc.) and the zeros mean that no non-zero element of $R(P)$ has there a non-zero entry. By the fact that $(x,y;\tilde{x},\tilde{y})$ satisfies M we have

(4.103) $$w_1 = -v(\tilde{y}_1)^*\,v(x_1) = v(\tilde{x}_1)^*\,v(y_1);\ w_2 = v(\tilde{y}_2)^*v(\tilde{x}_2) = -\,v(x_2)^*v(y_2)$$

(4.104) $$u_1 = v(y_1)v(x_1)^* = -v(\tilde{x}_1)v(\tilde{y}_1)^*;\quad u_2 = -v(\tilde{y}_2)v(\tilde{x}_2)^* = v(x_2)v(\tilde{y}_2)^*;$$

(4.105) $$\tilde{y}_1 = -x_1w_1^*,\quad y_1 = u_1x_1,\quad \tilde{x}_1 = u_1x_1w_1^*\ ;$$

(4.106) $$\tilde{y}_2 = \tilde{x}_2w_2^*,\quad y_2 = -u_2x_2,\quad x_2 = u_2\tilde{x}_2w_2^*\ .$$

Let $\tilde{P} = G(y)G(\tilde{y})P$. Then x and $\tilde{x}$ are atoms of $\tilde{P}$ and z, is an atom of $G(x)\tilde{P} = G(\tilde{x})\tilde{P} = Q$ such that $E(z)x = x$ and $E(z)\tilde{x} = x$. We can apply Lemma 4.2 to get

(4.107) $$|z_1| = |\tilde{x}_1|,\ |z_2| = |\tilde{x}_2|,\ |z_3| = |x_2|,\ |z_4| = |x_1|$$

(4.108) $$|z_1^*| = |x_2^*|,\ |z_2^*| = |x_1^*|,\ |z_3^*| = |\tilde{x}_1^*|,\ |z_4^*| = |\tilde{x}_2^*|$$

and

(4.109) $\tilde{x}_2 = \lambda v(z_4)\, x_1^* \, v(z_2); \quad \tilde{x}_1 = \lambda v(z_3)\, x_2^* \, v(z_1)$

for some $\lambda \in C$ with $|\lambda| = 1$.

Let $\theta \in C$ be any square root of $-\lambda$, and replace z by θz. Then (4.107) and (4.108) hold for the components of θz, and

(4.110) $\tilde{x}_2 = \lambda v(z_4)\, x_1^* \, v(z_2) = \bar{\theta}^2 \lambda v(\theta z_4)\, x_1^* \, v(\theta z_2) = -v(\theta z_4)\, x_1^* \, v(\theta z_2)$

(4.111) $\tilde{x}_1 = \bar{\theta}^2 \lambda v(\theta z_3)\, x_2^* v(\theta z_1) = -v(\theta z_3)\, x_2^* \, v(\theta z_1)$

Thus, there is no loss of generality to assume that z is chosen so that in (4.109) we have $\lambda = -1$, that is

(4.112) $\tilde{x}_2 = -v(z_4)\, x_1^* \, v(z_2) \; ; \quad \tilde{x}_1 = -v(z_3)\, x_2^* \, v(z_1)$.

Now, as one can easily check, z is the only atom of Q, up to a multiplication by a scalar. Thus, $E(x+y)z = E(x+\tilde{y})z = 0$. Using this we get as in Proposition 4.8 that:

(4.113) $z_3 = u_1 z_2 w_2^*, \quad z_1 = u_2 z_4 w_1^*$

Define $a_0 = x_1$ and

(4.114) $\sigma_1 = w_1, \; \sigma_2 = -v(z_2)^* v(x_1), \; \sigma_3 = w_2 \sigma_2 \;;$

(4.115) $\tau_1 = u_1, \; \tau_2 = v(z_4)\, v(x_1)^*, \; \tau_3 = u_2 \tau_2$.

Then, by (4.105), (4.106), (4.112) and (4.113) we get

(4.116) $y_1 = \tau_1 a_0, \; \tilde{y}_1 = -a_0 \sigma_1^*, \; \tilde{x}_1 = \tau_1 a_0 \sigma_1^*$

(4.117) $z_1 = \tau_3 a_0 \sigma_1^*, \; z_2 = -a_0 \sigma_2^*, \; z_3 = -\tau_1 a_0 \sigma_3^*, \; z_4 = \tau_2 a_0$

$$\tilde{x}_2 = -v(z_4)\, x_1^*\, v(z_2) = -\, v(z_4)v(x_1)^*\, x_1\, v(x_1)^* v(z_2) = \tau_2 a_0 \sigma_2^* \tag{4.118}$$

$$y_2 = -\tau_3 a_0 \sigma_2^*, \quad \tilde{y}_2 = \tau_2 a_0 \sigma_3^*, \quad x_2 = \tau_3 a_0 \sigma_3^* \tag{4.119}$$

This completes the proof in the first case, since formulas (4.116)-(4.119) are clearly equivalent to the existence of a tensor product representation in which formulas (4.93), (4.94) and (4.95) hold.

Next, assume that the second alternative holds, so we have (4.92). Now, y is an atom of $G(\tilde{x})G(x)P = \tilde{P}$ and z is an atom of $G(y)\tilde{P} = Q$. So, by Lemma 4.2, z is an atom of P. Also, $F(x)G(z)\tilde{x} = \tilde{x}$, so by Proposition 4.6 we get $F(z)G(x)P \neq 0$. Choose an atom $\tilde{z}$ of $F(z)G(x)P$ so that the system $(x,z;\tilde{x},\tilde{z})$ satisfies M. In particular, z is an atom of P, and thus an atom of Q. Also, (4.92) holds with $\tilde{z}$ instead of z. Thus, we have decompositions

$$x = \sum_{j=1}^{4} x_j, \quad y = \sum_{j=1}^{4} y_j, \quad z = \sum_{j=1}^{4} z_j, \quad \tilde{x} = \sum_{j=1}^{4} \tilde{x}_j, \quad \tilde{y} = \sum_{j=1}^{4} \tilde{y}_j, \quad \tilde{z} = \sum_{j=1}^{4} \tilde{z}_j \tag{4.120}$$

where the components $x_j, y_j, z_j, \tilde{x}_j, \tilde{y}_j$ and $\tilde{z}_j$ are defined by the following matrix:

(4.121)

	$\tilde{x}_2$	$\tilde{y}_2$	$\tilde{z}_2$				
x_1				$\tilde{y}_1$	$\tilde{z}_1$		
y_1				$\tilde{x}_1$		$\tilde{z}_4$	
z_1					$\tilde{x}_4$	$\tilde{y}_4$	
	y_2	x_2					$\tilde{z}_3$
	z_2		x_2				$\tilde{y}_3$
		z_3	y_3				$\tilde{x}_3$
				z_4	y_4	x_4	

where

(4.122) $|z_1| = |x_1| = |y_1|,\ |z_2| = |\tilde{x}_2| = |y_2|,\ |z_3| = |x_2| = |\tilde{y}_2|,$

$|z_4| = |\tilde{x}_1| = |y_1|,$

(4.123) $|\tilde{z}_1| = |\tilde{x}_4| = |y_4|,\ |\tilde{z}_2| = |x_3| = |y_3|,\ |\tilde{z}_3| = |\tilde{x}_3| = |\tilde{y}_3|,$

$|\tilde{z}_4| = |x_4| = |\tilde{y}_4|,$

(4.124) $|z_1| = |\tilde{x}_4^*| = |\tilde{y}_4^*|,\ |z_2^*| = |x_3^*| = |\tilde{y}_3^*|,\ |z_3^*| = |\tilde{x}_3^*| = |y_3^*|,$

$|z_4^*| = |x_4^*| = |y_4^*|,$

(4.125) $|\tilde{z}_1^*| = |x_1^*| = |\tilde{y}_1^*|,\ |\tilde{z}_2^*| = |\tilde{x}_2^*| = |\tilde{y}_2^*|,\ |\tilde{z}_3^*| = |x_2^*| = |y_2^*|,$

$|\tilde{z}_4^*| = |\tilde{x}_1^*| = |y_1^*|$

Consider first the following submatrix:

(4.126)

x_1	$\tilde{y}_1$	$\tilde{z}_1$	
y_1	$\tilde{x}_1$		$\tilde{z}_4$
z_1		$\tilde{x}_4$	$\tilde{y}_4$
	z_4	y_4	x_4

Since $(x,y;\tilde{x},\tilde{y})$ satisfies M we have

(4.127) $u_1 = v(y_1)\ v(x_1)^* = -v(\tilde{x}_1)v(\tilde{y}_1)^*;\ u = v(x_4)\ v(\tilde{y}_4)^* = -v(y_4)v(\tilde{x}_4)^*;$

(4.128) $w_1 = v(\tilde{x}_1)^* v(y_1) = -v(\tilde{y}_1)^* v(x_1)$; $w = v(\tilde{y}_4)^* v(\tilde{x}_4) = -v(x_4)^* v(y_4)$.

Similarly, since $(x,z;\tilde{x},\tilde{z})$ satisfies M, we have

(4.129) $u_2 = v(z_1)v(x_1)^* = -v(\tilde{x}_4)\, v(\tilde{z}_1)^*$

(4.130) $w_2 = v(\tilde{x}_4)^* v(z_1) = -v(\tilde{z}_1)^* v(x_1)$

Also,

(4.131) $E(x+y)\tilde{z} = 0 = E(x+\tilde{y})z$

and thus the same relations hold for the components described by (4.126). As in Proposition 4.8 we get from this

(4.132) $\tilde{z}_4 = u_1 \tilde{z}_1 w^*$, $z_4 = u z_1 w_1^*$

Put $a = x_1$ and define

(4.133) $w_3 = ww_2$, $u_3 = uu_2$

Then,

(4.134) $y_1 = u_1 a$, $\tilde{y}_1 = -aw_1^*$, $x_1 = u_1 a w_1^*$;

(4.135) $\tilde{z}_1 = -aw_2^*$, $\tilde{z}_4 = -u_1 a w_3^*$;

(4.136) $z_1 = u_2 a$, $z_4 = u_3 a w_1^*$;

(4.137) $\tilde{x}_4 = u_2 a w_2^*$, $\tilde{y}_4 = u_2 a w_3^*$

(4.138) $x_4 = u_3 a w_3^*$, $y_4 = -u_3 a w_2^*$

Clearly, the relations (4.134)-(4.138) can be written in an appropriate tensor product representation as

(4.139) $x_1 = e_{\{1\},\phi} \otimes a$; $x_4 = e_{\{1,2,3\},\{2,3\}} \otimes a$;

(4.140) $\tilde{x}_1 = e_{\{2\},\{1,2\}} \otimes a$; $\tilde{x}_4 = e_{\{3\},\{1,3\}} \otimes a$;

(4.141) $y_1 = e_{\{2\},\phi} \otimes a$; $y_4 = -e_{\{1,2,3\}\ \{1,3\}} \otimes a$;

(4.142) $\tilde{y}_1 = -e_{\{1\},\{1,2\}} \otimes a$; $\tilde{y}_4 = e_{\{3\},\{2,3\}} \otimes a$;;

(4.143) $z_1 = e_{\{3\},\phi} \otimes a$; $z_4 = e_{\{1,2,3\},\{1,2\}} \otimes a$;

(4.144) $\tilde{z}_1 = -e_{\{1\},\{1,3\}} \otimes a$; $\tilde{z}_4 = -e_{\{2\},\{2,3\}} \otimes a$.

Working in the same way with the components with the indices "2" and "3" we get that for some $b \in C_1$ with $\|b\|_1 + \|a\|_1 = 1$ the relations:

(4.145) $x_2 = e_{\{1,2\},\{2\}} \otimes b$; $x_3 = e_{\{1,3\},\{3\}} \otimes b$;

(4.146) $\tilde{x}_2 = e_{\phi,\{1\}} \otimes b$; $\tilde{x}_3 = e_{\{2,3\},\{1,2,3\}} \otimes b$;

(4.147) $y_2 = -e_{\{1,2\},\{1\}} \otimes b$; $y_3 = e_{\{2,3\},\{3\}} \otimes b$;

(4.148) $\tilde{y}_2 = e_{\phi,\{2\}} \otimes b$; $\tilde{y}_3 = -e_{\{1,3\},\{1,2,3\}} \otimes b$;

(4.149) $z_2 = -e_{\{1,3\},\{1\}} \otimes b$; $z_3 = e_{\{2,3\},\{2\}} \otimes b$;

(4.150) $\tilde{z}_2 = e_{\phi,\{3\}} \otimes b$; $\tilde{z}_3 = e_{\{1,2\},\{1,2,3\}} \otimes b$.

Summing up, formulas (4.134)-(4.150) yield the desired formulas (4.96), (4.97) and (4.98).

Finally, using formulas (4.97) and (4.98) and the fact that the system $(x_{3,2}, x_{3,3}; \tilde{x}_{3,2}, \tilde{x}_{3,3})$ satisfies M, we get that the system $(y,z,; \tilde{y},\tilde{z})$ satisfies M. □

5. CONSTRUCTION OF THE RANGE OF A CONTRACTIVE PROJECTION IN C_1 FROM ITS ATOMS, AND PROOFS OF THEOREMS 2.14, 2.15 AND 2.16.

(a) Reduction

In this section P will denote a given contractive projection in C_1. We show how $R(P)$ is constructed from its atoms, and thus prove Theorems 2.14 and 2.15. The proof of Theorem 2.16 follows then by duality.

We first introduce the notions of a "component" and of an irreducible subspace:

Definition 5.1: Let X be a subspace of C_1. A subspace Y of X is called a component of X if

(5.1) $E(Y)X = Y$ and $G(Y)X = \{0\}$

The trivial components of X are X and $\{0\}$. X is called irreducible if every component of X is trivial.

Proposition 5.2: Let X be a subspace of C_1. Then

(i) $X = \sum_{k\in K} X_k$, $1 \leq |K| \leq \aleph_0$ where $\{X_k\}_{k\in K}$ are pairwise disjointly supported subspaces of X, and thus in particular - components of X, and each subspace X_k is irreducible.

(ii) if P is a contractive projection from C_1 onto X, then $P_k = E(X_k)P = P\,E(X_k)P$ is a contractive projection from C_1 onto X_k and $Px = \sum_{k\in K} P_k x$ converges absolutely in C_1 for any $x \in C_1$.

Proof: (i) Let B_X denote the closed unit ball of X and let $\text{Ext}\, B_X = E_X$ denote the set of extreme points of B_X. By proposition 1.7, $B_X = \overline{\text{con}}\, E_X$ and $X = \overline{\text{span}}\, E_X$. If $X = \sum_{k=1}^{m} X_k$, $1 \leq m \leq \infty$, where the X_k's are pairwise disjointly supported subspaces of X, then it is easy to see that $E_X = \bigcup_{k=1}^{m} E_{X_k}$, and for $k \neq \ell$, $E_{X_k} \cap E_{X_\ell} = \phi$. If $x,y \in E_X$, define $x \sim y$ if y is contained in any component of X which contains x. This is clearly an equivalence relation which induces on E_X a partition: $E_X = \bigcup_{\alpha\in\Phi} E_\alpha$, $E_\alpha \cap E_\beta = \phi$ for $\alpha \neq \beta$, $E_\alpha \neq \phi$. Let $X_\alpha = \overline{\text{span}}\, E_\alpha$. If $\alpha \neq \beta$ and $x \in E_\alpha$ and $y \in E_\beta$, then $x \perp y$. This clearly implies that $X_\alpha \perp X_\beta$. Also, $E_{X_\alpha} = E_\alpha$ and thus X is irreducible, $\alpha \in \Phi$. Thus the family $\{X_\alpha\}_{\alpha\in\Phi}$ is actually countable, and we can rewrite it as $\{X_k\}_{k\in K}$ where $1 \leq |K| \leq \aleph_0$. Finally, $X = \sum_{k\in K} X_k$ follows easily from $X = \overline{\text{span}}\, E_X = \overline{\text{span}}\, (\bigcup_{k\in K} E_{X_k})$.

(ii) Since X_k is a subspace of X, the formula $E(X_k)P = P\,E(X_k)P$ is clear, and thus $P_k = E(X_k)P$ is a contractive projection from C_1 onto X_k. If $x \in C_1$, then $Px \in X = \sum_{k\in K} X_k$, and thus $Px = \sum_{k\in K} E(X_k)Px = \sum_{k\in K} P_k x$, the convergence (if $|K| = \aleph_0$) is in the norm of C_1, and it is actually an absolute convergence by the disjointness of the supports of the elements $P_k x$. □

Proposition 5.2 reduces our study of ranges of contractive projections in C_1 to the study of ranges of contractive projections in C_1 which are irreducible. Indeed, in order to prove Theorem 2.14 we have only to show that if X is an irreducible subspace of C_1 and X is the range of a contractive projection from C_1, then X is an elementary subspace of C_1. In our proof we will get more information which will enable us to prove also Theorem 2.15. We first prove a proposition which allows us to classify the irreducible ranges of contractive projections in C_1 into disjoint classes.

<u>Proposition 5.3:</u> Let P be a contractive projection in C_1 and let x be an atom of P. Then one and only one of the following mutually exclusive possibilities occurs:

<u>Case 1:</u> either $G(x)P = 0$, or there exists an atom y of $G(x)P$ so that $E(y)x = x$;

<u>Case 2:</u> there exists an atom y of P with $x\,G\,y$ and $F(y)G(x)P = 0$;

<u>Case 3:</u> there exist atoms $y,\tilde{y},\tilde{x}$ of P so that the system $(x,y;\tilde{x},\tilde{y})$ satisfies the matrix condition, and $G(x + \tilde{x})P = 0$;

<u>Case 4:</u> there exist atoms $y,\tilde{y},\tilde{x}$ of P as in case 3, and an atom z of either $G(y)G(x)G(x+\tilde{x})P$ or $G(\tilde{y})G(x)G(x+\tilde{x})P$, so that $F(z)G(\tilde{x})G(y)G(x)P \neq 0 \neq G(z)G(\tilde{x})G(y)G(x)P$.

<u>Case 5:</u> there exist atoms $y,\tilde{y},\tilde{x}$ and z of P as in case 4, and $F(z)G(\tilde{x})G(y)G(x)P = 0 = G(z)G(\tilde{x})G(y)G(x)P$.

<u>Proof:</u> If case 1 does not occur, then $G(x)P \neq 0$. If y is an atom of $G(x)P$, then by Lemma 4.2 we have $x\,G\,y$. If there exists an atom y of $G(x)P$ with $F(y)G(x)P = 0$ we are in case 2. If not, let y be any atom of $G(x)P$. By proposition 4.6 there exist atoms $\tilde{x}$ and $\tilde{y}$ of P so that the system $(x,y;\tilde{x},\tilde{y})$ satisfies the matrix condition. If we can choose $\tilde{x}$ and $\tilde{y}$ so that $G(x+\tilde{x})P = 0$, we are in case 3. If not, let $\tilde{x}$ and $\tilde{y}$ be as above, but assume that $G(x+\tilde{x})P \neq 0$. By the discussion before Proposition 4.9, either $G(y)G(x)G(x+\tilde{x})P \neq 0$ or o $G(\tilde{y})G(x)G(x+\tilde{x})P \neq 0$ (or both). Let z be an atom of one of these projections. By Proposition 4.9, if one of $F(z)G(\tilde{x})G(y)G(x)P$ and $G(z)G(\tilde{x})G(y)G(x)P$ is zero, then so is the other. Thus, if z can be chosen so that these projections are not zero, we are in case 4, and if for every such z these projections are both zero, we are in case 5. □

As we shall see latter, if $X = R(P)$ is irreducible and if we begin with an arbitrary atom x of P then:

(i) in case 1 X is an elementary subspace of C_1 of type 2, that is $X = SY_1^m(x)$;

(ii) in case 2 X is an elementary subspace of C_1 of type 4, and so $X = H_1^n(y_1,\dots,y_n)$;

(iii) in case 3 X is an elementary subspace of C_1 either of type 5 or of type 6, and thus either $X = AH_1^n(a,b)$ or $X = DAH_1^n(a)$ respectively;

(iv) in case 4 X is an elementary subspace of C_1 of type 3, so $X = A_1^m(b)$;

(v) in case 5 X is an elementary subspace of C_1 of type 1, so $X = C_1^{n,m}(a,b)$.

We assume now that P is a contractive projection in C_1 and that $X = R(P)$ is irreducible, and we treat each of the cases 1-5 described in proposition 5.3 separately.

<u>(b) Case 1 - Symmetric matrices - $SY_1^m(x)$</u>

<u>Lemma 5.4:</u> Let P be a contractive projection in C_1 so that $X = R(P)$ is irreducible, let x be an atom of P and let y be an atom of $G(x)P$ with $E(y)x = x$. Then there exist partial isometries $\{w_j\}_{1\le j\le m}$ and $\{u_j\}_{1\le j\le m}$ $(m \le \infty)$, with

(5.2) $w_1 = r(x)$, $u_1 = \ell(x)$; $r(w_j) = r(x)$, $r(u_j) = \ell(x)$ for $2 \le j$;

(5.3) for $i \ne j$ we have $\ell(w_j)\,\ell(w_i) = 0 = \ell(u_j)\,\ell(u_i)$;

and if we define

(5.4) $y_{i,j} = 2^{-1}(u_i\, x\, w_j^* + u_j\, x\, w_i^*)$; $1 \le i < j \le m$

(5.5) $y_{i,i} = u_i\, x\, w_i^*$; $1 \le i \le m$

then each $y_{i,i}$ is an atom of P, for $1 \le i < j \le m$ $y_{i,j}$ is an atom of both $G(y_{i,i})P$ and $G(y_{j,j})P$, and $X = \overline{\mathrm{span}}\,\{y_{i,j}\}_{1\le i\le j\le m}$.

Moreover if $a \in C_1$ and $F(X)a = 0$ then

(5.6) $Pa = \sum\limits_{1\le i\le j\le m} \langle a, v(y_{i,j})\rangle\, y_{i,j}$

Proof: Put $y_{1,1} = x$ and $y_{1,2} = y$. Let $\{y_{1,j}\}_{2\leq j\leq m}$, $2 \leq m \leq \infty$, be a maximal family of atoms of $G(x)P$ so that if $j_1 \neq j_2$ then $y_{1,j_1} G\, y_{1,j_2}$. By Lemma 4.2 and Remark 4.3 we have $E(y_{1,j})x = x$ for every j. Put $w_1 = r(x)$, $u_1 = \ell(x)$ and for $2 \leq j \leq m$ define:

$$(5.7) \qquad w_j = v(y_{1,j})^* v(x), \quad u_j = v(y_{1,j})v(x)^*$$

Then $\{w_j\}_{j=1}^m$ and $\{u_j\}_{j=1}^m$ are partial isometries which satisfy (5.2) and (5.3). By Lemma 4.2 we have for every $2 \leq j \leq m$:

$$(5.8) \qquad y_{1,j} = 2^{-1}\cdot(xw_j^* + u_j x)$$

and that

$$(5.9) \qquad y_{j,j} = u_j\, x\, w_j^* \text{ is an atom of } P.$$

If $1 \leq i < j \leq m$, then by Proposition 3.6(ii) we have that

$$(5.10) \qquad y_{i,j} = 2^{-1}(u_i x w_j^* + u_j x w_i^*)$$

belongs to X. Also, $G(y_{i,i})\, y_{i,j} = y_{i,j} = G(y_{j,j})y_{i,j}$ and $E(y_{i,j})y_{i,i} = y_{i,i}$ and $E(y_{i,j}y_{j,j} = y_{j,j}$, and this clearly implies that $y_{i,j}$ is an atom of both $G(y_{i,i})P$ and $G(y_{j,j})P$.

Put $y = \sum_{1\leq j\leq m} 2^{-j}\, y_{j,j}$. If $a \in R(E(y)P)$, then

$$(5.11) \qquad a = E(y)a = \sum_{1\leq j\leq m} E(y_{j,j})a + \sum_{1\leq i<j\leq m} E(y_{i,j})G(y_{i,i})a$$

$$= \sum_{1\leq j\leq m} \langle a, P^* v(y_{j,j})\rangle\, y_{j,j} + \sum_{1\leq i<j\leq m} \langle a, P^* v(y_{i,j})\rangle\, y_{i,j}$$

$$= \sum_{1\leq i\leq j\leq m} \langle a, P^* v(y_{i,j})\rangle\, y_{i,j}$$

Thus, $E(y)X = \overline{\text{span}}\, \{y_{i,j}\}_{1\leq i\leq j\leq m}$.

If for some $1 \leq j_0 \leq m$ $G(y_{j_0,j_0})\, G(y)P \neq 0$, let a be an atom of $G(y_{j_0,j_0})G(y)P$. If $j_0 = 1$, then $a \, G \, y_{1,j}$ for every $2 \leq j \leq m$ contradicts the maximality of $\{y_{1,j}\}_{2\leq j\leq m}$. If $j_0 \neq 1$, then by Remark 4.3 $E(a)y_{j_0,j_0} = y_{j_0,j_0}$, and thus by Lemma 3.6(ii) and Lemma 4.2, we get $G(a)G(y_{1,1})P \neq 0$, and this gives $G(y_{1,1})G(y)P \neq 0$ which is again a contradiction. This gives us

$$G(y)P = 0 \tag{5.12}$$

Since X is irreducible, its subspace $F(y)X$ is the zero subspace. Thus $E(y)X = X = \overline{\text{span}}\, \{y_{i,j}\}_{1\leq i\leq j\leq m}$.

It remains to prove (5.6). If $a \in C_1$ and $F(X)a = 0$, then by the computation (5.11) applied to Pa we have

$$Pa = \sum_{1\leq i\leq j\leq m} \langle a, (G(X) + E(X))\, P^* v(y_{i,j})\rangle\, y_{i,j} \, . \tag{5.13}$$

So it is enough to show that for every $1 \leq i \leq j \leq m$, $(G(X) + E(X))P^* v(y_{i,j}) = v(y_{i,j})$. This however is the consequence of Proposition 1.5 and the following proposition:

<u>Proposition 5.5:</u> Let $y_1, y_2 \in C_1$ with $y_1 \perp y_2$ and $\|y_1\|_1 = \|y_2\|_1 = 1$. Let $\tilde{P}$ be any contractive projection from C_1 onto $Y = \text{span}\{y_1, y_2\}$. Then $P^* v(y_1) \perp v(y_2)$ and $\tilde{P}^* v(y_2) \perp v(y_1)$.

Let us first show how to complete the proof of Lemma 5.4 using Proposition 5.5. Take $1 \leq i \leq j \leq m$ and put $y_1 = y_{i,j}$, $\tilde{y}_2 = \sum_{\substack{1\leq k\leq m \\ j\neq k\neq i}} 2^{-k}\, y_{k,k}$ and $y_2 = \tilde{y}_2 / \|\tilde{y}_2\|_1$. For $a \in C_1$ put $\tilde{P}a = \langle a, P^* v(y_1)\rangle\, y_1 + \langle a, P^* v(y_2)\, y_2$. Then $\tilde{P}$ is a contractive projection from C_1 onto $Y = \text{span}\{y_1, y_2\}$, with $\tilde{P}^* v(y_j) = P^*(y_j)$, $j = 1,2$. By Proposition 5.5 we have $P^* v(y_1) \perp v(y_2)$. Using Proposition 1.5 we get $P^* v(y_1) = v(y_1) + z$ where $z \in B(\ell_2)$, $\|z\|_\infty \leq 1$ and $z_1 \perp v(y_1)$. This clearly implies that $(G(X) + E(X))P^* v(y_{i,j}) = v(y_{i,j})$, completing the proof of Lemma 5.4.

Proof of Proposition 5.5: By symmetry it is enough to show that $\tilde{P}^* v(y_1) \perp y_2$. By proposition 1.5 we have $\tilde{P}^* v(y_1) = v(y_1) + z_1$, where $z_1 \in B(\ell_2)$, $\|z_1\|_\infty \leq 1$ and $z_1 \perp v(y_1)$. Now, one can easily verify that

$$(5.14) \qquad \|s\tilde{P}^* v(y_1) + t\tilde{P}^* v(y_2)\|_\infty = \max\{|s|,|t|\}$$

for every scalar s, t. Choose a matrix representation in which

$$(5.15) \qquad v(y_2) = \left[\begin{array}{c|c|c} 0 & 0 & 0 \\ \hline 0 & \begin{matrix} 1 & & 0 \\ & \ddots & \\ 0 & & 1 \; \ddots \end{matrix} & 0 \\ \hline 0 & 0 & 0 \end{array}\right]; \quad P^* v(y_1) = \left[\begin{array}{c|c|c} \begin{matrix} 1 & & 0 \\ & \ddots & \\ 0 & & 1 \; \ddots \end{matrix} & 0 & 0 \\ \hline 0 & z_{1,1} & z_{1,2} \\ \hline 0 & z_{2,1} & z_{2,2} \end{array}\right]$$

For $|s| \leq 1$ we have $1 = \|s\tilde{P}^* v(y_1) + \tilde{P}^* v(y_2)\|_\infty \geq \|s\, z_{1,1} + v(y_2)\|_\infty$. By [6], $v(y_2)$ is an extreme point of the unit ball of $E(v(y_2))B(\ell_2)$, and thus $z_{1,1} = 0$. Also, for $|s| \leq 1$

$$(5.16) \qquad 1 \geq \|(s\tilde{P}^* v(y_1) + \tilde{P}^* v(y_2)) r(y_2)\|_\infty^2 = \|s z_{2,1} + v(y_2)\|_\infty^2$$

$$= \|(s z_{2,1} + v(y_2))^* (s z_{2,1} + v(y_2))\|_\infty = \| |s|^2 |z_{2,1}|^2 + r(y_2)\|_\infty$$

$$= 1 + |s|^2 \|z_{2,1}\|_\infty^2$$

and thus $z_{2,1} = 0$. In the same way, $z_{1,2} = 0$, and so $z \perp v(y_2)$. Since we are given that $v(y_1) \perp v(y_2)$, we deduce that $(v(y_1)+z) \perp v(y_2)$. □

Remark 5.6: In tensor-product notations the conclusion of Lemma 5.4 can be written as $y_{i,j} = s_{i,j}/\|s_{i,j}\|_1 \otimes x$, where $s_{i,j}$ are the elementary symmetric matrices, $1 \leq i \leq j \leq m$. So $X = SY_1^m(x)$, $PG(X) = 0$ and $PE(X)$ is the canonical projection form C_1 onto X defined by formula (2.8).

(c) Case 5 - $C_1^{n,m}$ (a,b)

Lemma 5.7: Let P be a contractive projection in C_1 so that $X = R(P)$ is irreducible, let x be an atom of P and assume that case 5 of Proposition 5.3 occurs. Then there exist partial isometries $\{u_i\}_{i=1}^n$, $\{\sigma_i\}_{i=1}^n$, $\{w_j\}_{j=1}^m$ and $\{\tau_j\}_{j=1}^m$, where $2 \leq n \leq m \leq \infty$ and $3 \leq m$, and there exist $a,b \in C_1$ with a b and $\|a+b\|_1 = 1$, so that

(5.17) $r(u_i) = \ell(a)$, $r(w_j) = r(a)$; $r(\sigma_i) = r(b)$, $r(\tau_j) = \ell(b)$;

(5.18) $u_1 = \ell(a)$, $w_1 = r(a)$; $\sigma_1 = r(b)$, $\tau_1 = \ell(b)$;

(5.19) $\{\ell(w_j)\}_{j=1}^m \cup \{\ell(\sigma_i)\}_{i=1}^n$ are pairwise orthogonal;

(5.20) $\{\ell(u_i)\}_{i=1}^n \cup \{\ell(\tau_j)\}_{j=1}^m$ are pairwise orthogonal,

and so that each of

(5.21) $y_{i,j} = u_i a w_j^* + \tau_j b \sigma_{i=}^*$; $1 \leq i \leq n$, $1 \leq j \leq m$

is an atom of P, and $X = \overline{\text{span}}\{y_{i,j}$; $1 \leq i \leq n$, $1 \leq j \leq m$.

Moreover, if $c \in C_1$ and $F(X)c = 0$ then

$$(5.22) \qquad Pc = \sum_{\substack{1 \leq i \leq n \\ 1 \leq j \leq m}} \langle c, v(y_{i,j})\rangle\, y_{i,j}$$

Proof: Let $y, \tilde{y}, \tilde{x}$ and z be as in case 5 of Proposition 5.3, and assume for example that z is an atom of $G(y)G(x)G(x+\tilde{x})P$ (the proof in the other case is the same, since y and $\tilde{y}$ play symmetric roles). Note that by the assumptions in case 5 we have

$$(5.23) \qquad G(y)G(\tilde{x})G(x)P = 0$$

Clearly, $x \,G\, z$ and $y \,G\, z$ and $E(z)E(x)P = 0$. By Proposition 3.6 we must have also $E(z)F(x)P = 0$. It follows that z is actually an atom of P.

Put $a = \ell(z)\ell(y)x$, $b = x\,r(y)r(z)$, $w_1 = r(a)$, $u_1 = \ell(a)$, $\sigma_1 = r(b)$, $\tau_1 = \ell(b)$ and also:

$$(5.24)\qquad u_2 = v(\tilde{y})v(x)^* \;;\quad w_2 = v(y)^*v(x),\quad w_3 = v(z)^*v(x);$$

$$(5.25)\qquad \tau_2 = v(y)v(x)^*,\quad \tau_3 = v(z)v(x)^*;\quad \sigma_2 = v(\tilde{y})^*v(x).$$

By Lemma 4.2, $z = a\,w_3^* + \tau_3 b$, and using Proposition 4.6 we get that $\tilde{z} = u_2\, a\, w_3^* + \tau_3 b \sigma_2^*$ is an atom of P. Using in Proposition 4.9 the vectors $\{x,z,\tilde{z},\tilde{y},y\}$ instead of $\{x,y,\tilde{x},\tilde{y},z\}$, and the fact that $G(y)G(x+\tilde{z})P = 0$, we get $F(y)G(x+\tilde{z})P = 0$, and thus

$$(5.26)\qquad G(z)G(\tilde{y})G(x)P = 0$$

Put $y_{1,1} = x$, $y_{1,2} = y$, $y_{1,3} = z$, $y_{2,1} = \tilde{y}$, $y_{2,2} = \tilde{x}$, $y_{2,3} = \tilde{z}$. For short we denote $E(y_{i,j})$, $F(y_{i,j})$ and $G(y_{i,j})$ simply by $E_{i,j}$, $F_{i,j}$ and $G_{i,j}$ respectively. Choose now maximal families $\{y_{1,j}\}_{j=4}^{m}$ and $\{y_{i,1}\}_{i=3}^{n}$, $m,n \leq \infty$, of atoms of $G_{1,3}G_{1,2}G_{1,1}P$ and $G_{2,1}G_{1,1}P$ respectively, so that if $j_1 \neq j_2$ and $i_1 \neq i_2$ then

$$(5.27)\qquad y_{1,j_1} \,G\, y_{1,j_2},\quad y_{i_1,1} \,G\, y_{i_2,1}$$

It can be easily verified that (5.27) holds for any $1 \leq j_1, j_2 \leq m$ and $1 \leq i_1, i_2 \leq n$ with $j_1 \neq j_2$ and $i_1 \neq i_2$. Also, our analysis at the beginning of the proof shows that $y_{1,j}$ and $y_{i,1}$ are actually atoms of P, and that if we define for $4 \leq j$ and $i \geq 3$

$$(5.28)\qquad w_j = v(y_{1,j})^*v(y_{1,1}) \;;\quad \tau_j = v(y_{1,j})v(y_{1,1})^*$$

$$(5.29)\qquad u_i = v(y_{i,1})v(y_{1,1})^* \;;\quad \sigma_i = v(y_{i,1})^*v(y_{1,1})$$

then (5.17), (5.18), (5.19) and (5.20) hold, and (5.21) holds for pairs of the form $(i,1)$ and $(1,j)$. This last fact and Proposition 4.6 imply that if we define $y_{i,j}$ for any $1 \leq i \leq n$ and $1 \leq j \leq m$ by formula (5.21), then $y_{i,j}$ is an atom of P.

We claim now that if $1 \leq i_1, i_2 \leq n$, $1 \leq j_1, j_2 \leq m$, $i_1 \neq i_2$ and $j_1 \neq j_2$ then

$$(5.30) \qquad G_{i_2,j_1} G_{i_1,j_2} G_{i_1,j_1} P = 0.$$

Indeed, it is enough to prove (5.30) in two cases: for $i_1 = 1$ and $i_2 = 2$ and for arbitrary j_1, j_2, and for $j_1 = 1$, $j_2 = 2$ and arbitrary i_1 and i_2. This is done exactly as we get before (5.26) using proposition 4.9.

Put

$$(5.31) \qquad r_1 = \sum_{j=1}^{m} r(y_{1,j}),\ r_2 = \sum_{i=1}^{n} r(y_{i,1}),\ \ell_1 = \sum_{i=1}^{n} \ell(y_{i,1}),$$

$$\ell_2 = \sum_{j=1}^{m} \ell(y_{1,j})$$

then by (5.30) we get for any $c \in C_1$

$$(5.32) \qquad \ell_2 \cdot Pc \cdot r_1 + \ell_1 \cdot Pc \cdot r_2 = 0.$$

Put $\ell = \ell_1 + \ell_2$ and $r = r_1 + r_2$. We claim that for every $c \in C_1$:

$$(5.33) \qquad \ell . Pc \cdot (1-r) + (1-\ell) \cdot Pc \cdot r = 0$$

If not, then there exists either $1 \leq i_0 \leq n$ or $1 \leq j_0 \leq m$ so that either $(\prod_{j=1}^{m} G_{i_0,j})P \neq 0$ or $(\prod_{i=1}^{n} G_{i,j_0})P \neq 0$ respectively. If $i_0 = 1$ in the first case we get a contradiction to the maximality of $\{y_{1,j}\}_{j=4}^{m}$ by choosing an atom of $(\prod_{j=1}^{m} G_{1,j})P$. If $i_0 > 1$ in the first case, then using $y_{i_0,1}$, $y_{1,1}$, and any atom of $(\prod_{j=1}^{m} G_{i_0,j})P$ in proposition 4.6, we get that

$(\prod_{j=1}^{m} G_{1,j})P \neq 0$ which is again a contradiction. The proof in the second case is the same.

If $c = \ell \cdot c \cdot r \in R(P) = X$, then by the atomicity of the $y_{i,j}$

$$(5.34) \qquad c = \sum_{\substack{1\le i\le n \\ 1\le j\le m}} E(y_{i,j})c = \sum_{\substack{1\le j\le n \\ 1\le j\le m}} \langle c, P^* v(y_{i,j})\rangle \; y_{i,j}$$

and thus

$$(5.35) \qquad \ell \cdot X \cdot r = \overline{\text{span}}\, \{y_{i,j};\ 1 \le i \le n,\ 1 \le j \le m\} .$$

Using (5.33) and (5.35), we get that $(1-\ell)\cdot X\cdot(1-r)$ is a subspace of X, which by the irreducibility of X must be the zero subspace. This implies that $\ell = \ell(X)$, $r = r(X)$ and $X = \overline{\text{span}}\, \{y_{i,j};\ 1 \le i \le n,\ 1 \le j \le m\}$.

It remains to prove (5.22). If $c \in C_1$ and $F(X)c = 0$, then the computation (5.34) applied to Pc yields

$$(5.36) \qquad Pc = \sum_{\substack{1\le i\le n \\ 1\le j\le m}} \langle c,\ (G(X) + E(X))P^* \ v(y_{i,j})\rangle \ y_{i,j}$$

Thus, it is enough to prove that $(G(X) + E(X))P^* \ v(y_{i,j}) = v(y_{i,j})$ for every $1 \le i \le n$ and $1 \le j \le m$. Using proposition 5.5 we get that for any $1 \le i_1,i_2 \le n$, $1 \le j_1,j_2 \le m$ with $j_1 \neq j_2$

$$(5.37) \qquad P^* v(y_{i_1,j_1}) \perp v(y_{i_2,j_2})$$

As in the proof of Lemma 5.4, this implies that for any $1 \le i \le n$, $1 \le j \le m$ we have $(G(X) + E(X))P^* \ v(y_{i,j}) = v(y_{i,j})$.

Finally, we can assume without loss of generality that $n \le m$, and this completes the proof of Lemma 5.7. □

<u>Remark 5.8:</u> In tensor product notations the conclusions of Lemma 5.7 can be written as $y_{i,j} = e_{i,j} \otimes a + e_{j,i} \otimes b$, $1 \le i \le n$, $i \le j \le m$. So

$X = C_1^{n,m}(a,b)$, $PG(X) = 0$ and $PE(X)$ is the canonical projection of C_1 onto X defined by the analogous of formula (2.6).

(d) Case 4 - Antisymmetric matrices - $A_1^m(b)$

Lemma 5.9: Let P be a contractive projection in C_1 so that $X = R(P)$ is irreducible, let x be an atom of P and assume that case 4 of Proposition 5.3 occurs. Then there exists $a \in C_1$ with $\|a\|_1 = 1/2$ and there exist partial isometries $\{u_i\}_{i=1}^m$, $\{w_i\}_{i=1}^m$, $5 \leq m \leq \infty$, with

$$(5.38) \qquad r(u_i) = u_1 = \ell(a), \quad r(w_i) = w_1 = r(a); \quad 1 \leq i \leq m;$$

and

$$(5.39) \qquad \ell(u_i)\,\ell(u_j) = 0 = \ell(w_i)\,\ell(w_j); \quad \text{for } i \neq j, \quad 1 \leq i,j \leq m;$$

so that each of

$$(5.40) \qquad y_{i,j} = u_i\, a\, w_j^* - u_j\, a\, w_i^*, \quad 1 \leq i < j \leq m$$

is an atom of P, $x = y_{1,2}$, and $X = \overline{\text{span}}\,\{y_{i,j};\ 1 \leq i < j \leq m\}$.

Moreover, if $c \in C_1$ and $F(X)c = 0$ then

$$(5.41) \qquad Pc = \sum_{1 \leq i < j \leq m} \langle c, v(y_{i,j})\rangle\, y_{i,j}$$

Proof: By Lemma 4.9, there exists $a \in C_1$, $\|a\|_1 = 1/2$, and partial isometries $\{u_i\}_{i=1}^5$, $\{w_i\}_{i=1}^5$ so that (5.38), (5.39) and (5.40) are satisfied with 5 instead of m, and so that $x = y_{1,2}$. Let us denote for short $E_{i,j}$, $F_{i,j}$ and $G_{i,j}$ instead of $E(y_{i,j})$, $F(y_{i,j})$ and $G(y_{i,j})$ respectively. We choose a maximal family $\{y_{i,j}\}_{j=6}^m$, $5 \leq m \leq \infty$, of atoms of $G_{1,5}\, G_{1,4}\, G_{1,3}\, G_{1,2}\, P$ so that for $j_1 \neq j_2$, $y_{1,j_1}\ G\ y_{1,j_2}$. Clearly, this relation **holds** for any $1 \leq j_1, j_2 \leq m$ with $j_1 \neq j_2$. Also, $E_{1,j}\, E_{1,2}P = 0$, and thus by proposition 3.6 we have $E_{1,j}\, F_{1,2}P = 0$, which imply that $y_{1,j}$ is an atom of P for any $1 \leq j \leq m$. Put for $6 \leq j \leq m$

(5.42) $\quad u_j = v(y_{1,j})\, v(y_{1,2})^* u_2,\ w_j = v(y_{1,j})^* v(y_{1,2}) w_2$

By Lemma 4.2 and $y_{1,2}\ G\ y_{1,j}$ for $6 \leq j \leq m$, we get (5.40) with $i = 1$ and $2 < j \leq m$. Claim: for each $1 \leq i < j \leq m$, the $y_{i,j}$ defined by (5.40) is an atom of P. We prove this by induction on j. For $j = 2,3,4,5$ we know that $y_{i,j}$ is an atom of P if $1 \leq i < j$. Assume that $5 \leq j_0 < m$ and that each of y_{1,j_0}, $1 \leq i < j_0$, is an atom of P. Using y_{1,j_0}, y_{1,j_0+1} and y_{i,j_0} for $2 \leq i < j_0$, we get by Proposition 4.6 that y_{i,j_0+1} is an atom of P. Using $y_{1,2}$ and $(y_{1,j_0+1} + y_{2,j_0})/2$ in Proposition 3.6 we get that $y_{j_0,j_0+1} \in R(P)$. As can easily be verified, $y_{j_0,j_0+1}\ G\ y_{1,j_0+1}$ and $y_{j_0,j_0+1}\ G\ y_{2,j_0}$, and this implies by Lemma 4.2 that y_{j_0,j_0+1} is an atom of P. This completes the inductive proof.

We claim now that if $1 \leq i_0 \leq m$ then

(5.43) $\quad \left(\prod_{1\leq j<i_0} G_{j,i_0}\right)\cdot\left(\prod_{j_0<j\leq m} G_{i_0,j}\right)P = 0$

Indeed, for $i_0 = 1$ this follows by the maximality of the family $\{y_{1,j}\}_{j=6}^{m}$. If $i_0 > 1$, and the projection Q_{i_0} on the left hand side of (5.43) is not zero, we get by using in Proposition 4.6 y_{1,i_0+1}, y_{i_0,i_0+1} and an atom of Q_{i_0} if $i_0 < m$, or $y_{1,2}$, y_{2,i_0} and an atom of Q_{i_0} if $i_0 = m < \infty$, that $Q_1 \neq 0$ and this is a contradiction.

Put

(5.44) $\quad r = \sum_{j=1}^{m} \ell(w_j),\quad \ell = \sum_{j=1}^{m} \ell(u_j),$

and let $c = \ell \cdot c \cdot r \in R(P)$. By the atomicity of the $y_{i,j}$ we have that

(5.45) $\quad c = \sum_{1\leq i<j\leq m} E_{i,j}\, c = \sum_{1\leq i<j\leq m} \langle c, P^* v(y_{i,j})\rangle\, y_{i,j}$

so $\ell \cdot X \cdot r = \overline{\text{span}}\,\{y_{i,j};\ 1 \leq i < j \leq m\}$. As in the proofs of the previous cases, this implies using the irreducibility of X that $(1-\ell)\cdot X\cdot(1-r) = \{0\}$, and since by (5.43) $(1-\ell)\cdot X\cdot r = \{0\} = \ell\cdot X\cdot(1-r)$, we finally get

$\ell = \ell(X)$, $r = r(X)$ and $X = \overline{\text{span}}\,\{y_{i,j};\ 1 \leq i < j \leq m\}$.

It remains to prove (5.41). Let $c \in C_1$ with $F(X)c = 0$. Then by (5.45) we have

$$(5.46)\qquad Pc = \sum_{1 \leq i<j \leq m} \langle c, (G(X) + E(X))\, P^* v(y_{i,j})\rangle\, y_{i,j}$$

and thus it is enough to prove that $(G(X) + E(X))P^* v(y_{i,j}) = v(y_{i,j})$ for every $1 \leq i < j \leq m$. This is done by using Proposition 5.5 exactly as in the end of the proof of Lemma 5.4 by taking $y_1 = y_{i,j}$ but

$$\tilde{y}_2 = \sum_{\substack{1 \leq k<m \\ k \notin \{i-1,i,j-1,j\}}} 2^{-k}\, y_{k,k+1}\ .$$

□

<u>Remark 5.10:</u> In tensor product notations the conclusions of Lemma 5.9 can be written as $y_{i,j} = a_{i,j} \otimes a$, $1 \leq i < j \leq m$, where $a_{i,j}$ are the elementary anti-symmetric matrices. So $X = A_1^m(b)$, $b = a/\|a\|_1 = 2a$, $PG(X) = 0$, and $PE(X)$ is the canonical projection from C_1 onto $A_1^m(b)$ defined by a formula analogous to (2.9).

<u>(e)</u> <u>Case 2 - elementary Hilbert subspace</u> - $H_1^n(y_1,\ldots,y_n)$

This case is more complicated than the previous ones, and thus we divide the investigation in this case into three steps. We assume that P is a contractive projection in C_1 so that $X = R(P)$ is irreducible, and that for some atom x of P case 2 of Proposition 5.3 occurs, and our aim is to show that X is an elementary subspace of C_1 of type 4, that is a subspace of the form $H_1^n(y_1,\ldots,y_n)$ (see section 2b). In the first step we show that X is spanned by a sequence which is isometrically equivalent to the unit vector basis of $\ell_2^n (n \leq \infty)$, and any two members of it are G-related. In the second step we make a reduction to a case in which X has a simple form, and in the third step we prove that in this simple case $X = H_1^n(0,\ldots,0,y_m,0,\ldots,0)$ for some $1 \leq m \leq n$.

Lemma 5.11: Let P be a contractive projection in C_1 so that $X = R(P)$ is irreducible and assume that for some atom x of P case 2 of Proposition 5.3 occurs. Then, there exist a sequence $\{z_k\}_{k=1}^n$ of atoms of P with $z_1 = x$ so that $X = \overline{\text{span}}\ \{z_k\}_{k=1}^n$ and:

(5.47) $\quad z_{k_1}\ G\ z_{k_2}$ for $k_1 \neq k_2$;

(5.48) $\quad F(z_{k_1})\ G\ (z_{k_2})P = 0$ for $k_1 \neq k_2$;

(5.49) $\quad \|\sum_{k=1}^n t_k z_k\|_1 = (\sum_{k=1}^n |t_k|^2)^{1/2}$; for arbitrary scalars $\{t_k\}_{k=1}^n$

(5.50) each normalized element of X is an atom of P;

(5.51) if $a = \sum_{k=1}^n a_k z_k \neq 0 \neq \sum_{k=1}^n b_k z_k = b$, then aGb if and only if $\sum_{k=1}^n a_k \overline{b}_k = 0$;

(5.52) if $a \in C_1$ with $F(X)a = 0$, then $Pa = \sum_{k=1}^n \langle a, v(z_k)\rangle\, z_k$.

Proof: Let x,y be atoms of P with $x\ G\ y$ and $F(y)G(x)P = 0$. Put $z_1 = x$, $z_2 = y$, and let $\{z_k\}_{k=3}^n$, $2 \leq n \leq \infty$ be a maximal family of atoms of $G(z_2)G(z_1)P$ satisfying for $k_1 \neq k_2$ $z_{k_1}\ G\ z_{k_2}$. By Lemma 4.7, each z_k is an atom of P, (5.47) holds for any $k_1 \neq k_2$, and (5.48) holds for any $k_1 \neq k_2$ such that $\{k_1,k_2\} \cap \{1,2\} \neq \phi$. Let $3 \leq k_1,k_2 \leq n$ with $k_1 \neq k_2$, and apply Lemma 4.7 with $x = z_1$, $y = z_{k_1}$ and $z = z_{k_2}$ to get (5.48). So (5.48) holds for any $1 \leq k_1,k_2 \leq n$ with $k_1 \neq k_2$.

We claim now that for any finite $1 \leq m \leq n$, and for any $0 \neq a = \sum_{k=1}^m t_k z_k$ we have $\|a\|_1 = (\sum_{k=1}^m |t_k|^2)^{1/2}$, and that $a/\|a\|_1 = \tilde{a}$ is an atom of P which satisfies $\tilde{a}\ G\ z_k$ for every $m < k$.

If $m = 1$, or $m = 2$ this follows from Lemma 4.7 and the construction of the $\{z_k\}_{k=3}^n$. Assume that our claim holds for some m and let us prove it for $m + 1$. Let $0 \neq a = \sum_{k=1}^{m+1} t_k z_k$. If $t_{n+1} = 0$ or $b = \sum_{k=1}^{m} t_k z_k = 0$, we are trivially done. So assume that $t_{n+1} \neq 0$ and $b \neq 0$. By the induction hypothesis, $\|b\|_1 = (\sum_{k=1}^{m} |t_k|^2)^{1/2}$ and $b/\|b\|_1 = \tilde{b}$ is an atom of P which satisfies $\tilde{b}\ G\ z_{m+1}$. By Lemma 4.2 we get that

$$\|a\|_1 = \|\ \|b\|_1 \tilde{b} + t_{m+1} z_{m+1} \|_1 = (\|b\|_1^2 + |t_{m+1}|^2)^{1/2} = (\sum_{k=1}^{m+1} |t_k|^2)^{1/2},$$

and that $\tilde{a} = a/\|a\|_1$ is an atom of P. If $m+1 < k$, then $z_k\ G\ \tilde{b}$ and $z_k\ G\ z_{m+1}$. So, by Lemma 4.7, $z_k\ G\ \tilde{a}$. This completes the inductive proof of the claim above and thus establishes (5.49).

Denote for short E_k, F_k and G_k instead of $E(z_k)$, $F(z_k)$ and $G(z_k)$. By (5.47) and (5.48) we get for $k \neq k'$ that $G_k F_{k'} P = E_k F_{k'} P = 0$ and $E_k G_{k'} P = E_k P$, and thus for any finite $1 \leq m \leq n$ we get by induction

$$P = \sum_{k=1}^{m} E_k P + \prod_{k=1}^{m} F_k P + \prod_{k=1}^{m} G_k P \tag{5.53}$$

By the maximality of $\{z_k\}_{k=1}^n$ we clearly have $(\prod_{k=1}^{n} G_k)P = 0$. If $a \in C_1$ is arbitrary, we get by (5.53) (and letting $m \to \infty$ in case $n = \infty$) that

$$Pa = \sum_{k=1}^{n} E_k Pa + (\prod_{k=1}^{n} F_k)Pa = \sum_{k=1}^{n} \langle a, P^* v(z_k)\rangle z_k + (\prod_{k=1}^{n} F_k)Pa . \tag{5.54}$$

Put $r = \sup_{1 \leq k \leq n} r(z_k)$, $\ell = \sup_{1 \leq k \leq n} \ell(z_k)$. Then from (5.54), $\ell \cdot X \cdot r = \overline{\text{span}}\ \{z_k\}_{k=1}^n$ and $(1-\ell) \cdot X \cdot (1-r) = \{0\} = \ell \cdot X \cdot (1-r)$. Hence $(\prod_{k=1}^{n} F_k)X = (1-\ell) \cdot X \cdot (1-r)$ is a subspace of X, and since X is irreducible this subspace is zero. This implies that

$$r = r(X),\ \ell = \ell(X),\ X = \overline{\text{span}}\ \{z_k\}_{k=1}^n \tag{5.55}$$

(5.56) $$Pa = \sum_{k=1}^{n} \langle a, P^* v(z_k)\rangle z_k$$

From the fact that X is a Hilbert space (hence, every normalized element of X is an extreme point of the unit ball of X) we get (5.50) by using Proposition 3.3. To prove (5.51), we first notice that if $((\cdot,\cdot))$ is the inner product of X, then $((\sum_{k=1}^{n} a_k z_k, \sum_{k=1}^{n} b_k z_k)) = \sum_{k=1}^{n} a_k \bar{b}_k$. If $a,b \in X$ with $a \neq 0 \neq b$ and $a\ G\ b$, then by Lemma 4.2: $\|a+b\|_1^2 = \|a\|_1^2 + \|b\|_1^2$ and so $((a,b)) = 0$. Conversely, if $((a,b)) = 0$ and $\|a\|_1 = 1 = \|b\|_1$, we denote $Y = \text{span}\{a,b\}$ and Q is the orthogonal projection of X onto Y. Thus $P_1 = QP$ is a contractive projection from C_1 onto Y. Clearly, a is an atom of P_1 and $F(a)P_1 = 0$ (since otherwise we get that Y contains ℓ_1^2). So $G(a)P_1 \neq 0$. Choose an atom c of $G(a)P_1$ and note that $c\ G\ a$, $((c,a)) = 0$, $Y = \text{span}\{a,c\}$. It follows that b is proportional to c and thus $b\ G\ a$.

It remains to prove (5.52), and this will follow from (5.56) and the following

<u>Proposition 5.12:</u> Let $x_1, x_2 \in C_1$

(5.57) $$\|t_1x_1 + t_2x_2\|_1 = (|t_1|^2 + |t_2|^2)^{1/2} \text{ for any scalars } t_j,$$

and assume that $\tilde{P}$ is a contractive projection from C_1 onto $\text{span}\{x_1,x_2\}$. Then

(5.58) $$(\tilde{P}^* v(x_1) - v(x_1)) \perp v(x_2)$$

We first complete the proof of Lemma 5.11 using Propositions 5.12. Let $a \in C_1$ with $F(X)a = 0$, then by (5.56) we get

(5.59) $$Pa = \sum_{k=1}^{n} \langle a, (G(X) + E(X))P^* v(z_k)\rangle z_k$$

and thus (5.52) will be the consequence of $(G(X) + E(X))P^* v(z_k) = v(z_k)$, $1 \leq k \leq n$. Let $k_1 \neq k_2$ and put $x_j = z_{k_j}, j = 1,2$. Let Q be the orthogonal contractive projection from X onto $\text{span}\{x_1,x_2\}$ and let $P = QP$ be the induced contractive projection from C_1 onto $\text{span}\{x_1,x_2\}$. Using Proposition 1.5 we get for $j = 1,2$

$$(5.60)\qquad \tilde{P}^* v(z_{k_j}) = P^* v(z_{k_j}) = v(z_{k_j}) + a_j,\ \|a_j\|_\infty \leq 1,\ \ a_j \perp v(z_{k_j})$$

Since, by proposition 5.12, $a_1 \perp v(z_{k_2})$, we get actually $F(z_{k_1})F(z_{k_2})a_1 = a_1$, and since k_2 is arbitrary, we get

$$(5.61)\qquad (\prod_{k=1}^{n} F_k)a_1 = a_1$$

and thus

$$(5.62)\qquad (G(X) + E(X))a_1 = 0 \quad\text{and}\quad (G(X) + E(X))P^* v(z_{k_1}) = v(z_{k_1}).$$

Since k_1 is arbitrary, this completes the proof of Lemma 5.11.

Proof of Proposition 5.12: We show first that $x_1\ G\ x_2$. Let y_1 be any atom of $\tilde{P}$, then $F(y_1)\tilde{P} = 0$ (since $R(\tilde{P})$ does not contain ℓ_1^2), and thus there exists an atom y_2 of $G(y_1)\tilde{P}$. Now $R(\tilde{P})$ is two-dimensional, and thus by Lemma 4.2 we get $y_1\ G\ y_2$ and y_2 is an atom of $\tilde{P}$. We are in case 2 of Proposition 5.3 (with y_1, y_2 and $\tilde{P}$ instead of x, y and P), so from (5.51) and (5.57) we get $x_1\ G\ x_2$. Note that we can apply the conclusions of the second case in Lemma 4.2 to x_1, x_2.

By Proposition 1.5 we have for $j = 1,2$

$$(5.63)\qquad \tilde{P}^* v(x_j) = v(x_j) + a_j,\ \ \|a_j\|_\infty \leq 1,\ \ a_j \perp v(x_j).$$

Also, clearly, for every scalar t_1, t_2:

$$(5.64)\qquad \|t_1\tilde{P}^* v(x_1) + t_2\tilde{P}^* v(x_2)\|_\infty = (|t_1|^2+|t_2|^2)^{1/2}$$

Let us choose a matrix representation in which

$$(5.65)\quad v(x_1) = \begin{pmatrix} 0 & v_1^{(1)} & 0 & 0 \\ v_1^{(2)} & 0 & 0 & 0 \\ 0 & 0 & 0 & 0 \\ 0 & 0 & 0 & 0 \end{pmatrix} \quad v(x_2) = \begin{pmatrix} 0 & 0 & v_2^{(1)} & 0 \\ 0 & 0 & 0 & 0 \\ v_2^{(2)} & 0 & 0 & 0 \\ 0 & 0 & 0 & 0 \end{pmatrix}$$

and each of the $v_j^{(i)}$, $1 \leq i,j \leq 2$, is diagonal with only "1" on its diagonal. In this representation we get by (5.63)

$$(5.66)\quad \tilde{P}^* v(x_1) = \begin{pmatrix} 0 & v_1^{(1)} & 0 & 0 \\ v_1^{(2)} & 0 & 0 & 0 \\ 0 & 0 & a_1^{(1)} & a_1^{(2)} \\ 0 & 0 & a_1^{(3)} & a_1^{(4)} \end{pmatrix} \quad \tilde{P}^* v(x_2) = \begin{pmatrix} 0 & 0 & v_2^{(1)} & 0 \\ 0 & a_2^{(1)} & 0 & a_2^{(2)} \\ v_2^{(2)} & 0 & 0 & 0 \\ 0 & a_2^{(3)} & 0 & a_2^{(4)} \end{pmatrix}$$

where $\{a_j^{(i)}\}_{i=1}^4$ are the components of a_j, $j = 1,2$.
We want to show that $a_j^{(i)} = 0$, except perhaps if $i = 4$. This is the consequence of (5.64), (5.66) and the following:

Claim: Let α,β be complex numbers and assume that for every complex number λ we have $(1 + |\lambda|^2)^{1/2} \geq \|\begin{pmatrix} 1 & \lambda \\ \lambda\alpha & \beta \end{pmatrix}\|_\infty$. Then $\alpha = \beta = 0$.

Proof: For any 2×2 matrix A we have

$$(5.67)\qquad 2\,\|A\|_\infty^2 = \text{trace}\, A^*A + ((\text{trace}\, A^*A)^2 - 4 \det A^*A)^{1/2}$$

Take $\lambda \in C$ so that $\lambda^2 = \text{sgn}(\bar{\alpha}\beta)$ if $\alpha\beta \neq 0$, and $\lambda = 1$ otherwise. Then by (5.67) we get

(5.68) $$4 \geq 2 \left\| \begin{pmatrix} 1 & \lambda \\ \lambda\alpha & \beta \end{pmatrix} \right\|_\infty^2 = 2 + |\alpha|^2 + |\beta|^2 + [4 + (|\alpha|^2 + |\beta|^2)^2 + 8|\alpha|\,|\beta|]^{1/2}$$

and this implies that $\alpha = \beta = 0$.

This completes the proof of Proposition 5.12. □

Definition 5.13: A sequence $\{a_k\}$ of non-zero elements in C_1 is said to satisfy the strong G-condition if

(5.69) $$a_k \; G \; y \quad \text{for every } k \text{ and every } 0 \neq y \in \operatorname{span}\{a_j\}_{1 \leq j < k} \;;$$

(5.70) $$\left\| \sum_k t_k a_k \right\|_1 = \left(\sum_k |t_k|^2 \right)^{1/2} \|a_1\|_1; \quad \text{for every scalar } \{t_k\}\,.$$

For short we say that $\{a_k\}$ satisfy SG. Note that (5.70) implies that $\|a_k\|_1 = \|a_1\|_1$ for every k.

Let us continue our analysis in case 2 of Proposition 5.3, and let P, X and $\{z_k\}_{k=1}^n$ be as in Lemma 5.11. Clearly, $\{z_k\}_{k=1}^n$ satisfy SG. In the sequel we shall use only the fact that $\{z_k\}_{k=1}^n$ satisfy SG in order to prove that in an appropriate tensor product representation we have $z_k = h_{n,k}^{(1)}(y_1,\ldots,y_n)$, in the sense of formula (2.32), for some disjointly supported elements $y_m \in C_1$. We assume from now on that $n < \infty$, and in the end of this subsection we shall consider the case $n = \infty$ which is much simpler.

Notations: $I_n = \{1,2,\ldots,n\}$, $\mathcal{P}_n$ denotes the set of all subsets of I_n. By α,β,γ we denote elements of $\mathcal{P}_n$. If $1 \leq k,m \leq n$, then we define

(5.71) $$\mathcal{P}_n^k = \{\alpha \in \mathcal{P}_n;\ k \in \alpha\}; \qquad \mathcal{P}_{n,m} = \{\alpha \in \mathcal{P}_n;\ |\alpha| = m\}$$

where $|\alpha|$ denotes the number of elements of α. We put also $\mathcal{P}_{n,m}^k = \mathcal{P}_{n,m} \cap \mathcal{P}_n^k$, $1 \leq k,m \leq n$.

For $1 \leq k \leq n$ and $\alpha \in \mathscr{P}_n^k$ we put

$$(5.72)\qquad z_{k,\alpha} = \Big(\prod_{\substack{i\in\alpha\sim\{k\}\\ j\in I_n\sim\alpha}} \ell(z_i)(1-\ell(z_j))\Big) z_k \Big(\prod_{\substack{i\in\alpha\sim\{k\}\\ j\in I_n\sim\alpha}} r(z_j)(1-r(z_i))\Big)$$

and finally, for $1 \leq k,m \leq n$ we put

$$(5.73)\qquad z_{k,m} = \sum_{\alpha\in\mathscr{P}_{n,m}^k} z_{k,\alpha} .$$

Note that $z_{k,\alpha}$ and $z_{k,m}$ depend on the whole sequence $\{z_j\}_{j=1}^n$ and not only on k,α,m.

Proposition 5.14: Let $\{z_k\}_{k=1}^n$ be a normalized sequence in C_1 which satisfies SG. Then

(5.74) if $1 \leq k \leq n$ and $\alpha,\beta \in \mathscr{P}_n^k$ with $\alpha \neq \beta$, then $z_{k,\alpha} \perp z_{k,\beta}$;

$$(5.75)\qquad z_k = \sum_{\alpha\in\mathscr{P}_n^k} z_{k,\alpha} = \sum_{m=1}^n z_{k,m}, \quad 1 \leq k \leq n;$$

$$(5.76)\qquad 1 = \sum_{\alpha\in\mathscr{P}_n^k} \|z_{k,\alpha}\|_1 = \sum_{m=1}^n \|z_{k,m}\|_1, \quad 1 \leq k \leq n;$$

$$(5.77)\qquad \ell(z_{k,\alpha}) = \Big(\prod_{i\in\alpha} \ell(z_i)\Big)\cdot\Big(\prod_{j\in I_n\sim\alpha} (1-\ell(z_j))\Big), \quad 1 \leq k \leq n, \ \alpha \in \mathscr{P}_n^k ;$$

$$(5.78)\qquad r(z_{k,\alpha}) = \Big(\prod_{j\in(I_n\sim\alpha)\cup\{k\}} r(z_j)\Big)\cdot\Big(\prod_{i\in\alpha\sim\{k\}} (1-r(z_i))\Big), \quad 1 \leq k \leq n, \ \alpha \in \mathscr{P}_n^k ;$$

(5.79) if $1 \leq k_1,k_2 \leq n$, and $1 \leq m_1 \neq m_2 \leq n$ then $z_{k_1,m_1} \perp z_{k_2,m_2}$;

(5.80) for $1 \leq m \leq n$, either $z_{k,m} = 0$ for every $1 \leq k \leq n$, or the sequence $\{z_{k,m}\}_{k=1}^n$ satisfies SG.

Proof: formulas (5.74)-(5.78) are just easy consequences of the definitions (5.72) and (5.73) and the fact that $\{z_k\}_{k=1}^n$ satisfies SG. If

$1 \leq k_1, k_2 \leq n$ and $1 \leq m_1 \neq m_2 \leq n$, then for $\alpha \in \mathcal{P}^{k_1}_{n,m_1}$ and $\beta \in \mathcal{P}^{k_2}_{n,m_2}$ we have by (5.77) and (5.78) $z_{k_1,\alpha} \perp z_{k_2,\beta}$. This fact and (5.73) gives us (5.79).

If $2 \leq k \leq n$, $1 \leq m \leq n$, $y \in \operatorname{span}\{z_{j,m}\}_{j=1}^{k-1}$ and $y \neq 0 \neq z_{k,m}$ then $y \mathrel{G} z_{k,m}$ follows from (5.79) and the SG condition for $\{z_j\}_{j=1}^n$. This last fact, formula (1.8) and [1, Proposition 3.1] imply that for every scalar $\{t_k\}_{k=1}^n$ we have for every $1 \leq m \leq n$;

$$(5.81) \qquad \Big\| \sum_{k=1}^n t_k z_{k,m} \Big\|_1 \geq \Big(\sum_{k=1}^n |t_k|^2 \, \| z_{k,m} \|_1^2 \Big)^{1/2} .$$

Thus for every scalar $\{t_k\}_{k=1}^n$ we get:

$$(5.82) \qquad \Big(\sum_{k=1}^n |t_k|^2 \Big)^{1/2} = \Big\| \sum_{m=1}^n \sum_{k=1}^n t_k z_{k,m} \Big\|_1 =$$

$$= \sum_{m=1}^n \Big\| \sum_{k=1}^n t_k z_{k,m} \Big\|_1 \geq \sum_{m=1}^n \Big(\sum_{k=1}^n |t_k|^2 \, \| z_{k,m} \|_1^2 \Big)^{1/2} \geq$$

$$\geq \Big(\sum_{k=1}^n \Big(\sum_{m=1}^n |t_k| \, \| z_{k,m} \|_1 \Big)^2 \Big)^{1/2}$$

$$= \Big(\sum_{k=1}^n |t_k|^2 \Big(\sum_{m=1}^n \| z_{k,m} \|_1 \Big)^2 \Big)^{1/2} = \Big(\sum_{k=1}^n |t_k|^2 \Big)^{1/2}$$

Thus we have an equality everywhere in (5.82), and so for each $1 \leq m \leq n$ we have an equality in (5.81). If $t_k = 1$ for every $1 \leq k \leq n$, then the equality in (5.82) gives us that for some numbers $0 \leq \lambda_k$, μ_m, $\| z_{k,m} \|_1 = \lambda_k \mu_m$. If $k_1 \neq k_2$ then $\| z_{k_1,m} \|_1 \lambda_{k_2} = \| z_{k_2,m} \|_1 \lambda_{k_1}$ for every m. Summing on m and using (5.76) we get that $\lambda_{k_2} = \lambda_{k_1}$, and thus we can rewrite $\| z_{k,m} \|_1 = \mu_m$, $1 \leq k, m \leq n$. This completes the proof of (5.80), and thus completes the proof of the Proposition. $\square$

Definition 5.15: A sequence $\{a_k\}_{k=1}^n$ of non-zero elements in C_1 is said to satisfy the strong G-condition of type m, $1 \leq m \leq n$, in notation SG_m, if it satisfies SG and if $a_k = a_{k,m}$ for every $1 \leq k \leq n$, where $a_{k,m}$ is defined by formulas (5.72) and (5.73) in which a_j replaces z_j for $1 \leq j \leq n$.

By Proposition 5.14 and the SG condition for the sequence $\{z_k\}_{k=1}^n$ we have for every $1 \leq m \leq n$ such that $z_{k,m} \neq 0$ that the sequence $\{z_{k,m}\}_{k=1}^n$ satisfies SG_m. In order to show that in an appropriate tensor product representation we have $z_k = h_{n,k}^{(1)}(y_1,\ldots,y_n)$ for some disjointly supported $y_m \in C_1$, it is enough by view of (5.79) to prove that for fixed $1 \leq m \leq n$, $z_{k,m} = x_{n,k,m} \otimes y_m$, $1 \leq k \leq n$, where the $x_{n,k,m}$ were defined by (2.15). In order to avoid complicated notations we assume simply that the sequence $\{z_k\}_{k=1}^n$ itself satisfies SG_m for some $1 \leq m \leq n$, and that $\|z_k\|_1 = 1$ for $1 \leq k \leq n$.

Proposition 5.16: Let $\{z_k\}_{k=1}^n$ be a sequence of normalized elements of C_1 which satisfies SG_m for some $1 \leq m \leq n$. Then for every $1 \leq j,k \leq n$:

(i) if $\alpha \in \mathscr{P}_{n,m}^k \cap \mathscr{P}_{n,m}^j$ then $|z_{k,\alpha}^*| = |z_{j,\alpha}^*|$;

(ii) if $\alpha \in \mathscr{P}_{n,m}^k$, $\beta \in \mathscr{P}_{n,m}^j$ with $\alpha = (\beta \sim \{j\}) \cup \{k\}$ then $|z_{k,\alpha}| = |z_{j,\beta}|$;

(iii) if $\alpha \in \mathscr{P}_{n,m}^k$, $\beta \in \mathscr{P}_{n,m}^j$ then there exist partial isometries u and w so that $r(u) = \ell(z_{k,\alpha})$, $\ell(u) = \ell(z_{j,\beta})$, $r(w) = r(z_{k,\alpha})$ and $\ell(w) = r(z_{j,\beta})$, and $z_{j,\beta} = u\, z_{k,\alpha}\, w^*$;

(iv) if $\alpha \in \mathscr{P}_{n,m}^k$ then $\|z_{k,\alpha}\|_1 \cdot \binom{n-1}{m-1} = 1$.

Proof: (i) Let $k \neq j$, then z_k G z_j. If $\alpha \in \mathscr{P}_{n,m}^k \cap \mathscr{P}_{n,m}^j$ then $\ell(z_{k,\alpha}) = \ell(z_{j,\alpha})$ by (5.77). By Proposition 1.2 and (5.74), we get $|z_{k,\alpha}^*| = |z_{j,\alpha}^*|$.

(ii) Here we must have $k \neq j$, and by (5.78) $r(z_{k,\alpha}) = r(z_{j,\beta})$. So $|z_{k,\alpha}| = |z_{j,\beta}|$ follows by the same reasons as in (i).

(iii) If $\alpha = \beta$, then by (i) we get

$$(5.83) \quad z_{j,\beta} = |z^*_{j,\beta}|\ v(z_{j,\beta}) = |z^*_{k,\alpha}|\ v(z_{j,\alpha}) = z_{k,\alpha}\ v(z_{k,\alpha})^* v(z_{j,\alpha}).$$

So our claim holds with $u = \ell(z_{k,\alpha}) = \ell(z_{j,\alpha})$ and $w = v(z_{j,\alpha})^* v(z_{k,\alpha})$.

If $|\alpha \sim \beta| = 1 = |\beta \sim \alpha|$, let $\{k_1\} = \alpha \sim \beta$ and $\{j_1\} = \beta \sim \alpha$. By the case $\alpha = \beta$ treated before we get $z_{j,\beta} = u_1\ z_{j_1,\beta}\ w_1^*$ and $z_{k_1,\alpha} = u_2\ z_{k,\alpha}\ w_2^*$ (where u_1, u_2, w_1 and w_2 are partial isometries which satisfy the requirements in (iii). Also, since $\alpha = (\beta \sim \{j_1\}) \cup \{k_1\}$ we get by (ii) that $|z_{j_1,\beta}| = |z_{k_1,\alpha}|$ and thus $z_{j_1,\beta} = u_3\ z_{k_1,\alpha}\ w_3^*$. Thus, summing up, we have

$$(5.84) \quad z_{j,\beta} = u\ z_{k,\alpha}\ w^*, \quad u = u_1 u_3 u_2, \quad w = w_1 w_3 w_2$$

where u,w are partial isometries with $r(u) = \ell(z_{k,\alpha})$, $\ell(u) = \ell(z_{j,\beta})$, $r(w) = r(z_{k,\alpha})$ and $\ell(w) = r(z_{j,\beta})$.

If α, β are arbitrary with $\alpha \in \mathscr{P}^k_{n,m}$ and $\beta \in \mathscr{P}^j_{n,m}$, then there exists a sequence $\alpha = \alpha_1, \alpha_2, \ldots, \alpha_\nu = \beta$ $(1 \leq \nu \leq n)$ so that $|\alpha \sim \beta| = 1 = |\beta \sim \alpha|$ we get partial isometries u_i, w_i with $r(u_i) = \ell(z_{k_i,\alpha_i})$, $\ell(u_i) = \ell(z_{k_{i+1},\alpha_{i+1}})$, $r(w_i) = r(z_{k_i,\alpha_i})$, $\ell(w_i) = r(z_{k_{i+1},\alpha_{i+1}})$, and so that $z_{k_{i+1},\alpha_{i+1}} = u_i\ z_{k_i,\alpha_i}\ w_i^*$. It follows that $z_{j,\beta} = u\ z_{k,\alpha}\ w^*$, with $u = u_{\nu-1} \cdot u_{\nu-2} \cdots u_1$ and $w = w_{\nu-1} \cdot w_{\nu-2} \cdot \cdots \cdot w_1$. Clearly, the partial isometries u and w satisfy $r(u) = \ell(z_{k,\alpha}), \ell(u) = \ell(z_{j,\beta})$, $r(w) = r(z_{k,\alpha})$ and $\ell(w) = r(z_{j,\beta})$.

(iv) This is a trivial consequence of (iii), Proposition 5.14, and the fact that $\{z_k\}_{k=1}^n$ satisfies the condition SG_m. □

Lemma 5.17: Let $\{z_k\}_{k=1}^n$ be a normalized sequence in C_1 which satisfies SG_m for some $1 \leq m \leq n$. Then there exists an element $y \in C_1$ with $\|y\|_1 = 1/\binom{n-1}{m-1}$, and there exists a tensor product representation in which:

$$(5.85) \quad z_k = x_{n,k,m} \otimes y, \quad 1 \leq k \leq n$$

where the $x_{n,k,m}$ are the matrices defined by (2.15).

Proof: If $\alpha \in \mathscr{P}_{n,m}$ and $k \in \alpha$ is any element we put $\ell_\alpha = \ell(z_{k,\alpha})$. If $\beta \in \mathscr{P}_{n,m-1}$ and $k \in I_n \sim \beta$ we put $r_\beta = r(z_{k,\beta\cup\{k\}})$. By (5.77) and (5.78) these definitions are indpendent of the particular k. Clearly, for every $k \in I$ and $\alpha \in \mathscr{P}^k_{n,m}$ we have $z_{k,\alpha} = \ell_\alpha \cdot z_{k,\alpha} \cdot r_{\alpha \sim \{k\}}$. Also, if $\alpha,\alpha' \in \mathscr{P}_{n,m}$ and $\beta,\beta' \in \mathscr{P}_{n,m-1}$ with $\alpha \neq \alpha'$ and $\beta \neq \beta'$, then $\ell_\alpha \cdot \ell_{\alpha'} = 0 = r_\beta \cdot r_{\beta'}$. We therefore can regard each of the z_k as an operator matrix $z_k = ((z_k)_{\alpha,\beta})$, $\alpha \in \mathscr{P}_{n,m}$, $\beta \in \mathscr{P}_{n,m-1}$ with

$$(5.86) \quad (z_k)_{\alpha,\beta} = \begin{cases} z_{k,\alpha}, & \beta = \alpha \sim \{k\}, \quad \alpha \in \mathscr{P}^k_{n,m} \\ 0, & \text{otherwise} \end{cases}$$

Claim: Let $\alpha_m = \{1,2,3,\ldots,m\}$, $\beta_m = \alpha_m \sim \{m\} = \{1,2,\ldots,m-1\}$, and $y_m = (-1)^{m+1} z_{m,\alpha_m}$. Then there exist partial isometries $\{u_\alpha\}_{\alpha\in\mathscr{P}_{n,m}}$ and $\{w_\beta\}_{\beta\in\mathscr{P}_{n,m-1}}$ with $u_{\alpha_m} = \ell_{\alpha_m}$, $w_{\beta_m} = r_{\beta_m}$, so that for every $\alpha \in \mathscr{P}_{n,m}$ and $\beta \in \mathscr{P}_{n,m-1}$:

$$(5.87) \quad r(u_\alpha) = \ell_{\alpha_m}, \ \ell(u_\alpha) = \ell_\alpha \ ; \ r(w_\beta) = r_{\beta_m}, \ \ell(w_\beta) = r_\beta$$

and for every $1 \leq k \leq n$ and $\alpha \in \mathscr{P}^k_{n,m}$ we have

$$(5.88) \quad z_{k,\alpha} = (-1)^{i(\alpha,k)} \cdot u_\alpha y_m w^*_{\alpha \sim \{k\}}$$

where $i(\alpha,k) = |\{k' \in \alpha \ ; \ k' < k\}|$

It is clearly enough to prove the claim, since then for every $1 \leq k \leq n$ we get

$$(5.89) \qquad z_k = \sum_{\alpha \in \mathscr{P}^k_{n,m}} z_{k,\alpha} = \sum_{\alpha \in \mathscr{P}^k_{n,m}} (-1)^{i(\alpha,k)} \cdot u_\alpha y_m w^*_{\alpha \sim \{k\}}$$

which is clearly equivalent to the existence of a tensor product representation in which

$$(5.90) \qquad z_k = \sum_{\alpha \in \mathscr{P}^k_{n,m}} (-1)^{i(\alpha,k)} e_{\alpha,\alpha \sim \{k\}} \otimes y_m = x_{n,k,m} \otimes y_m$$

We prove the claim in the following inductive manner: we first prove it for an arbitrary natural number n, and $m = 1$ or $m = n$. Then we assume that $1 < m < n$ and that the claim is true for the pairs $(n-1,m-1)$ and $(n-1,m)$ and prove it for the pair (n,m).

If $m = 1$ and n is arbitrary, then for every $1 \leq k \leq n$ we get

$$(5.91) \qquad z_k = z_{k,\{k\}} = (\prod_{j \neq k} (1-\ell(z_j)))\, z_k\, (\prod_{j \neq k} r(z_j))$$

and by (ii) of Proposition 5.16:

$$(5.92) \qquad |z_k| = |z_j|, \quad \text{for every } 1 \leq j,k \leq n.$$

Put $y_1 = (-1)^{i(\{1\},1)} \cdot z_{1,\{1\}} = z_1$, and define for $1 \leq k \leq n$

$$(5.93) \qquad u_{\{k\}} = v(z_k)v(z_1)^*, \quad w_\phi = r_\phi = r(z_1)$$

Then (5.87) and (5.88) are consequences of (5.91), (5.92) and the definition (5.93). Note that this proof works also for $n = \infty$ and $m = 1$.

If $m = n$, then $\alpha_n = \{1,2,\ldots,n\}$ and for every $1 \leq k \leq n$ we get

$$(5.94) \qquad z_k = z_{k,\alpha_n} = (\prod_{j \neq k} \ell(z_j))\, z_k\, (\prod_{j \neq k} (1 - r(z_j)))$$

and by (i) of Proposition 5.16:

(5.95) $\quad |z_k^*| = |z_j^*|$, for every $1 \leq j,k \leq n$.

Put $y_n = (-1)^{n+1} z_n = (-1)^{n+1} z_{n,\alpha_n}$, $u_{\alpha_n} = \ell_{\alpha_n} = \ell(z_n)$

and for $1 \leq k \leq n$ define

(5.96) $$w_{\alpha_n \sim \{k\}} = (-1)^{n-k} v(z_k)^* v(z_n)$$

Again, (5.87) and (5.88) follow from (5.94), (5.95) and (5.96). We prove for example (5.88): if $1 \leq k \leq n$ then $\mathcal{P}^k_{n,m} = \{\alpha_n\}$ and so

(5.97) $$z_{k,\alpha_n} = z_k = |z_k^*| v(z_k) = |z_n^*| v(z_k) = z_n v(z_n)^* v(z_k)$$

$$= (-1)^{n+1} y_n w^*_{\alpha_n \sim \{k\}} \cdot (-1)^{-n+k} = (-1)^{k+1} u_{\alpha_n} y_n w^*_{\alpha_n \sim \{k\}} .$$

Note again that the proof in this case works also for $m = n = \infty$.

Let us proceed by induction and assume that $2 < n < \infty$ and $1 < m < n$, and that the claim is true for the pairs $(n-1,m-1)$ and $(n-1,m)$. For every $1 \leq k < n$ we can write

(5.98) $$z_k = \tilde{z}_k + \tilde{\tilde{z}}_k, \text{ where } \tilde{\tilde{z}}_k = \sum_{\substack{\alpha \in \mathcal{P}^k_{n,m} \\ n \notin \alpha}} z_{k,\alpha}, \quad \tilde{z}_k = \sum_{\substack{\alpha \in \mathcal{P}^k_{n,m} \\ n \in \alpha}} z_{k,\alpha}$$

Clearly, $\tilde{z}_k \perp \tilde{\tilde{z}}_j$ for every $1 \leq j,k < n$. Also, the proof of (5.80) of Proposition 5.14 works also in our case, and we get that $\{z_k\}_{k=1}^{n-1}$ satisfies SG_{m-1} and that $\{\tilde{\tilde{z}}_k\}_{k=1}^{n-1}$ satisfies SG_m. Clearly $\|\tilde{z}_k\|_1 = \frac{m-1}{n-1}, \|\tilde{\tilde{z}}_k\|_1 = \frac{m-1}{n-1}$ for $1 \leq k < n$. Note also that if $\alpha \in \mathcal{P}^k_{n,m}$ $1 \leq k < n$, then $z_{k,\alpha} = \tilde{z}_{k,\alpha}$ if $n \in \alpha$, and $z_{k,\alpha} = \tilde{\tilde{z}}_{k,\alpha}$ if $n \notin \alpha$. By our induction hypothesis for the pair $(n-1,m)$ we get partial isometries $\{u_\alpha\}_{\alpha \in \mathcal{P}_{n-1,m}}$ and $\{w_\beta\}_{\beta \in \mathcal{P}_{n-1,m-1}}$ with $u_{\alpha_m} = \ell_{\alpha_m}$, $w_{\beta_m} = r_{\beta_m}$ and such that (5.87) and (5.88) hold with

$n-1$ instead of n, and with $y_m = (-1)^{m+1} \cdot z_{m,\alpha_m}$. Let $\gamma = (\alpha \sim \{m\}) \cup \{n\}$, then by (ii) of Proposition 5.16 $|z_{n,\gamma}| = |z_{m,\alpha_m}|$, and by (5.77) we have $\ell(z_{n,\gamma}) \cdot \ell(z_{m,\alpha_m}) = 0$. Define $u_\gamma = v(z_{n,\gamma}).v(z_{m,\alpha_m})^*$. Then, clearly, $z_{n,\gamma} = (-1)^{m+1} u_\gamma y_m$. Also, by (i) of Proposition 5.16 we have $|z^*_{n,\gamma}| = |z^*_{m-1,\gamma}|$, and by (5.78): $r(z_{n,\gamma}) \cdot r(z_{m-1,\gamma}) = 0$. Define $w_{\gamma\sim\{m-1\}} = -v(z_{m-1,\gamma})^* v(z_{n,\gamma})$, then $z_{m-1,\gamma} = (-1)^m u_\gamma y_m w^*_{\gamma\sim\{m-1\}}$. Note that the mapping $\alpha \to \alpha \cup \{n\}$ is a bijection of $\mathcal{P}_{n-1,m-1}$ onto $\mathcal{P}^n_{n,m}$ and of $\mathcal{P}_{n-1,m-2}$ onto $\mathcal{P}^n_{n,m-1}$. Thus by our induction hypothesis on the pair $(n-1,m-1)$, there exist partial isometries $\{\tilde{u}_\alpha\}_{\alpha\in\mathcal{P}^n_{n,m}}$ and $\{\tilde{w}_\beta\}_{\beta\in\mathcal{P}^n_{n,m-1}}$ with $r(\tilde{u}_\alpha) = \ell_\gamma$, $r(\tilde{w}_\beta) = r_{\gamma\sim\{m-1\}}$, $\tilde{u}_\gamma = \ell_\gamma$, $\tilde{w}_{\gamma\sim\{m-1\}} = r_{\gamma\sim\{m-1\}}$, $\ell(\tilde{u}_\alpha) = \ell_\alpha$, $\ell(\tilde{w}_\beta) = r_\beta$, and for $1 \le k \le n-1$ and $\alpha \in \mathcal{P}^k_{n,m}$ we have

$$(5.99) \qquad z_{k,\alpha} = \tilde{z}_{k,\alpha} = (-1)^{i(\alpha,k)} \tilde{u}_\alpha (-1)^m z_{m-1,\gamma} \tilde{w}^*_{\alpha \sim \{k\}}$$

$$= (-1)^{i(\alpha,k)} \tilde{u}_\alpha u_\gamma y_m w^*_{\gamma \sim \{m-1\}} \tilde{w}^*_{\alpha \sim \{k\}}$$

Put for $\alpha \in \mathcal{P}^n_{n,m}$ and $\beta \in \mathcal{P}^n_{n,m-1}$: $u_\alpha = \tilde{u}_\alpha u_\gamma$ and $w_\beta = \tilde{w}_\beta w_{\gamma \sim \{m-1\}}$. Clearly, (5.87) holds for every $\alpha \in \mathcal{P}_{n,m}$ and $\beta \in \mathcal{P}_{n,m-1}$, and (5.88) holds for every $1 \le k \le n-1$, and $\alpha \in \mathcal{P}^k_{n,m}$ and for $k = n$ and $\alpha = \gamma$. It remains to prove (5.88) for $k = n$ and $\gamma \ne \alpha \in \mathcal{P}^n_{n,m}$. To this end it is clearly enough to prove that if $\alpha,\beta \in \mathcal{P}^n_{n,m}$ with $|\alpha\sim\beta| = |\beta\sim\alpha| = 1$ and if $z_{n,\alpha} = (-1)^{m+1} u_\alpha y_m w^*_{\alpha\sim\{n\}}$ then also $z_{n,\beta} = (-1)^{m+1} u_\beta y_m w^*_{\beta\sim\{n\}}$.

So let $\alpha,\beta \in \mathcal{P}^n_{n,m}$, let $1 \le j,k \le n-1$ with $j \ne k$ and so that $k \in \alpha$, $j \in \beta$, $\alpha \sim \{k\} = \beta \sim \{j\}$ and assume that $z_{n,\alpha} = (-1)^{m+1} u_\alpha y_m w^*_{\alpha \sim \{n\}}$. Assume without loss of generality that $j < k$. Put

$$(5.100) \qquad \alpha' = (\alpha \sim \{n\}) \cup \{j\} = (\beta \sim \{n\}) \cup \{k\}$$

$$(5.101) \qquad \beta' = \alpha \sim \{k\} = \beta \sim \{j\}$$

$$(5.102) \qquad b = b_1 + b_2, \quad b_1 = z_{j,\alpha'}, \quad b_2 = z_{j,\beta}$$

$$(5.103) \qquad c = c_1 + c_2, \quad c_1 = z_{k,\alpha'}, \quad c_2 = z_{k,\alpha}$$

$$(5.104) \qquad d = d_1 + d_2, \quad d_1 = z_{n,\beta}, \quad d_2 = z_{n,\alpha}$$

Since $\{z_j, z_k, z_n\}$ satisfy SG, we get by the proof of (5.80) in Proposition 5.14 that $\{b,c,d\}$ satisfy SG_2. We can therefore apply Proposition 4.8 to these $\{b,c,d\}$. Put

$$(5.105) \qquad a = u_{\alpha'} y_m w^*_{\beta'}$$

and note that

$$(5.106) \qquad i(\alpha',k) = i(\alpha,k) + 1; \quad i(\alpha',j) = i(\beta,j).$$

By the induction hypothesis we get

$$(5.107) \qquad b_1 = z_{j,\alpha'} = a((-1)^{i(\beta,j)} w_{\beta'} \cdot w^*_\alpha \sim \{n\})$$

$$(5.108) \qquad b_2 = z_{j,\beta} = ((-1)^{i(\beta,j)} u_\beta u^*_{\alpha'})a$$

$$(5.109) \qquad c_1 = z_{k,\alpha'} = -a((-1)^{i(\alpha,k)} w_{\beta'} w^*_\beta \sim \{n\})$$

$$(5.110) \qquad c_2 = z_{k,\alpha} = ((-1)^{i(\alpha,k)} u_\alpha u^*_{\alpha'})a$$

If for example $i(\alpha,k) + i(\beta,j) \equiv m (\bmod 2)$ we get by our assumption on $z_{n,\alpha}$:

$$(5.111) \qquad z_{n,\alpha} = -((-1)^{i(\alpha,k)} u_\alpha u^*_{\alpha'})a((-1)^{i(\beta,j)} w_{\beta'} w^*_\alpha \sim \{n\}).$$

Thus, by proposition 4.8 we must have

$$(5.112)\qquad z_{n,\beta} = -((-1)^{i(\beta,j)} u_\beta\, u^*_{\alpha'})\, a((-1)^{i(\alpha,k)}\, w_{\beta'}\, w^*_{\beta\sim\{n\}})$$

$$= (-1)^{m-1}\, u_\beta\, y_m\, w^*_{\beta\sim\{n\}}$$

If $i(\alpha,k) + i(\beta,j) \equiv m+1 \pmod 2$, we get instead of (5.11) that

$$(5.113)\qquad z_{n,\alpha} = ((-1)^{i(\alpha,k)} u_\alpha\, u^*_{\alpha'})\, a((-1)^{i(\beta,j)}\, w_{\beta'}\, w^*_{\alpha\sim\{n\}})$$

and thus, by Proposition 4.8 we get

$$(5.114)\qquad z_{n,\beta} = ((-1)^{i(\beta,j)} u_\beta\, u^*_{\alpha'})\, a((-1)^{i(\alpha,k)}\, w_{\beta'}\, w^*_{\beta\sim\{n\}})$$

$$= (-1)^{m-1}\, u_\beta\, y_m\, w^*_{\alpha\sim\{n\}}\,.$$

This completes the proof of the claim, and thus also the proof of Lemma 5.17. □

<u>The case $n = \infty$:</u> Let us go back to the end of Lemma 5.11 and continue our analysis in the case $n = \infty$. We assume as before that the normalized sequence $\{z_k\}_{k=1}^\infty$ only satisfies SG. Let I be the set of all positive integers, and for $k \in I$ put $\mathcal{P}^k = \{\alpha \subseteq I; k \in \alpha\}$. If $\alpha \in \mathcal{P}^k$, we put

$$(5.15)\qquad z_{k,\alpha} = \Big(\prod_{\substack{i\in\alpha\sim\{k\}\\ j\in I\sim\alpha}} \ell(z_i)(1-\ell(z_j))\Big)\, z_k \Big(\prod_{\substack{i\in\alpha\sim\{k\}\\ j\in I\sim\alpha}} r(z_j)(1-r(z_i))\Big)$$

We can do all the analysis of Proposition 5.14 and Lemma 5.17 also in this case, but now for $1 < m < \infty$, $z_{k,m} = \sum_{\substack{|\alpha|=m\\ \alpha\in\mathcal{P}^k}} z_{k,\alpha}$ is the sum of infinitely many pairwise disjointly supported elements of equal norm, so $z_{k,m} = 0$. So, if $z_{k,\alpha} \neq 0$ then either $\alpha = \{k\}$, or α is an infinite set.

If $\alpha \in \mathcal{P}^k$ is infinite, $\alpha \neq I$ and $z_{k,\alpha} \neq 0$, then for every $j \in \alpha$ we get by Proposition 5.16(i) that $|z^*_{j,\alpha}| = |z^*_{k,\alpha}|$ and thus, in particular, $\|z_{j,\alpha}\|_1 = \|z_{k,\alpha}\|_1 \neq 0$. Also, by Proposition 5.16(ii),

$|z_{k',(\alpha\sim j)\cup\{k\}}| = |z_{j,\alpha}|$ for every $k' \in I \sim \alpha$, and thus $\|z_{k',(\alpha\sim j)\cup k'}\|_1 = \|z_{j,\alpha}\|_1 = \|z_{k,\alpha}\|_1 \neq 0$. Fix $k' \in I \sim \alpha$, and let for every $j \in \alpha$, $\alpha_j = (\alpha \sim j)\cup\{k'\}$. Then $\{z_{k',\alpha_j}\}_{j\in\alpha}$ are pairwise disjointly supported, and $\|z_{k',\alpha_j}\|_1 = \|z_{k,\alpha}\|_1 \neq 0$ for every $j \in \alpha$. But this leads to a contradiction since $\|z_{k'}\|_1 \geq \sum_{j\in\alpha} \|z_{k',\alpha_j}\|_1$, and since α is infinite.

It follows that for every $k \in I$,

$$(5.116) \qquad z_k = z_{k,\{k\}} + z_{k,I}$$

Of course, if $k,j \in I$, then $z_{k,\{k\}} \perp z_{j,I}$, and the sequences $\{z_{k,\{k\}}\}_{k=1}^{\infty}$ and $\{z_{k,I}\}_{k=1}^{\infty}$ satisfy SG, unless they are zero. As in the cases $m = 1$ and $m = n$ of Lemma 5.17, we get elements $y_1, y_\infty \in C_1$ with $y_1 \perp y_\infty$ and $\|y_1 + y_\infty\|_1 = 1$ and an appropriate tensor representation in which for every $k \in I$

$$(5.117) \qquad z_k = z_{\infty,k,1} \otimes y_1 + x_{\infty,k,\infty} \otimes y_\infty$$

where the matrices $x_{\infty,k,1}$ and $x_{\infty,k,\infty}$ were defined by (2.16). □

We sum the work in this subsection as follows

<u>Corollary 5.18:</u> Let $\{z_k\}_{k=1}^{n}$, $1 \leq n \leq \infty$, be a normalized sequence in C_1 which satisfies SG. If $n < \infty$, then there exist elements $y_1,\ldots,y_n \in C_1$ with $\sum_{m=1}^{n} \|y_m\|_1 \binom{n-1}{m-1} = 1$, and there exists a tensor product representation so that for every $1 \leq k \leq n$:

$$(5.118) \qquad z_k = h_{n,k}^{(1)}(y_1,\ldots,y_n) = \sum_{m=1}^{n} x_{n,k,m} \otimes y_m \, .$$

If $n = \infty$, then there exist $y_1, y_\infty \in C_1$ with $\|y_1\|_1 + \|y_\infty\|_1 = 1$, and there exists a tensor product representation in which for $1 \leq k < \infty$

(5.119) $z_k = h^{(1)}_{\infty,k}(y_1,y_\infty) = x_{\infty,k,1} \otimes y_1 + x_{\infty,k,\infty} \otimes y_\infty$.

In both cases there exists a canonical contractive projection from C_1 onto $\overline{\text{span}}\ \{z_k\}^n_{k=1}$, in the sense of subsection 2b.

In particular, if P is a contractive projection in C_1 with $X = R(P)$ irreducible, and if for some atom x of P case 2 of proposition 5.3 occurs, then X is an elementary subspace of C_1 of type 4, $PG(X) = 0$, and $PE(X)$ is the canonical projection from C_1 onto X.

Remark 5.19: Let X be an elementary subspace of type 4, and let $\{z_k\}^n_{k=1}$ ($n \leq \dim X$) be any orthonormal sequence in X (remember that in this case X is an Hilbert space). Then $\{z_k\}^n_{k=1}$ satisfies SG and $Z = [z_k]^n_{k=1}$ is the range of a contractive projection $\tilde{P}$ from C_1. So Corollary 5.18 with $\tilde{P}$ instead of P and Z instead of X is valid.

(f) - Case 3 - elementary subspaces of the forms $AH^n_1(a,\tilde{a})$ and $DAH^n_1(a)$

Recall that a system $(x_1,x_2;\tilde{x}_1,\tilde{x}_2)$ of normalized elements of C_1 satisfies the matrix condition (condition "M", for short, see Definition 4.4 and Proposition 4.5) if there exists a tensor product representation and elements $a,b \in C_1$ with $a \perp b$ and $\|a+b\|_1 = 1$ so that

(5.120) $x_1 = e_{1,1} \otimes a + e_{1,1} \otimes b;\quad x_2 = e_{1,2} \otimes a + e_{2,1} \otimes b;$

(5.121) $\tilde{x}_1 = e_{2,2} \otimes a + e_{2,2} \otimes b;\quad \tilde{x}_2 = -\ (e_{2,1} \otimes a + e_{1,2} \otimes b).$

We first introduce some conditions which are closely related to the matrix condition.

Definition 5.20: A system $(x,\tilde{x},y)$ of normalized elements in C_1 is said to satisfy the diagonal condition, (condition "D", for short) if there exist partial isometries u and w with $r(u) = \ell(x)$, $r(w) = r(x)$ and $r(u) \cdot \ell(u) = 0 = r(w) \cdot \ell(w)$, and so that

(5.122) $y = \frac{1}{2}(ux + xw^*)$, $\tilde{x} = -uxw^*$

It is clear that condition D for $(x,\tilde{x},y)$ is equivalent to the existence of a matrix representation in which

(5.123) $$x = \begin{pmatrix} a & 0 \\ 0 & 0 \end{pmatrix}; \quad \tilde{x} = \begin{pmatrix} 0 & 0 \\ 0 & -a \end{pmatrix}; \quad y = \frac{1}{2}\begin{pmatrix} 0 & a \\ a & 0 \end{pmatrix}$$

and also, to the existence of a tensor product representation in which

(5.124) $$x = e_{1,1} \otimes a; \quad \tilde{x} = -e_{2,2} \otimes a; \quad y = \frac{1}{2}(e_{1,2} + e_{2,1}) \otimes a,$$

where $\|a\|_1 = 1$.

It is also clear that if $(x,\tilde{x},y)$ satisfies D, then $(\tilde{x},x,y)$ satisfies D, and that if $(x_1,x_2;\tilde{x}_1,\tilde{x}_2)$ satisfies M, then $(x_1,\tilde{x}_1,\frac{1}{2}(x_2 + \tilde{x}_2)$ satisfies D.

Definition 5.21: (i) A system $(x_1,\ldots,x_n;\tilde{x}_1,\ldots,\tilde{x}_n)$, $2 \leq n < \infty$, of normalized elements in C_1 is said to satisfy the strong matrix condition (condition SM, for short), if for every $1 \leq j,k \leq n$ with $j \neq k$ $(x_j,x_k;\tilde{x}_j,\tilde{x}_k)$ satisfies the matrix condition

(ii) A system $(x_1,\ldots,x_n;\tilde{x}_1;\ldots,\tilde{x}_n;y)$ $1 \leq n \leq \infty$, of normalized elements in C_1 is said to satisfy the diagonal strong matrix condition (condition DSM, for short), if the system $(x_1,\ldots,x_n;\tilde{x}_1,\ldots,\tilde{x}_n)$ satisfies SM, and for every $1 \leq j \leq n$ the system $(x_j,\tilde{x}_j,y)$ satisfies D.

The following proposition gives us some information on systems satisfying condition M (and thus - information on systems satisfying SM or DSM), which will be needed in the sequel.

Proposition 5.22: Let $x_1,x_2,\tilde{x}_1$ and $\tilde{x}_2$ be normalized elements in C_1.

(i) Assume that $x_k = a_k + b_k$, $\tilde{x}_k = \tilde{a}_k + \tilde{b}_k$, $k = 1,2$, where $\mathrm{span}\{a_k,\tilde{a}_k\}_{k=1}^2 \perp \mathrm{span}\{b_k,\tilde{b}_k\}_{k=1}^2$. Then, $(x_1,x_2;\tilde{x}_1,\tilde{x}_2)$ satisfies the matrix condition if and only if there exist numbers $0 \leq \lambda,\mu \leq 1$ with $\lambda + \mu = 1$, so that

(*) $\|a_j\|_1 = \|\tilde{a}_j\|_1 = \lambda$, $\|b_j\|_1 = \|\tilde{b}_j\|_1 = \mu$, $j = 1,2$;

($\overset{*}{*}$) if $\lambda > 0$, then $(\lambda^{-1}a_1,\lambda^{-1}a_2;\lambda^{-1}\tilde{a}_1,\lambda^{-1}\tilde{a}_2)$ satisfies the matrix condition.

($\overset{**}{*}$) if $\mu > 0$, then $(\mu^{-1}b_1,\mu^{-1}b_2;\mu^{-1}\tilde{b}_1,\mu^{-1}\tilde{b}_2)$ satisfies the matrix condition

(ii) Assume that $(x_1,x_2;\tilde{x}_1,\tilde{x}_2)$ satisfies the matrix condition, and that there exists a matrix representation in which:

$$(5.125)\quad x_1 = \begin{pmatrix} x_{1,1} & 0 & 0 & 0 \\ 0 & 0 & 0 & 0 \\ 0 & 0 & x_{1,2} & 0 \\ 0 & 0 & 0 & 0 \end{pmatrix}; \quad x_2 = \begin{pmatrix} 0 & x_{2,1} & 0 & 0 \\ 0 & 0 & 0 & 0 \\ 0 & 0 & 0 & 0 \\ 0 & 0 & x_{2,2} & 0 \end{pmatrix}$$

$$(5.126)\quad \tilde{x}_1 = \begin{pmatrix} 0 & 0 & 0 & 0 \\ 0 & \tilde{x}_{1,1} & 0 & 0 \\ 0 & 0 & 0 & 0 \\ 0 & 0 & 0 & \tilde{x}_{1,2} \end{pmatrix}; \quad \tilde{x}_2 = \begin{pmatrix} 0 & 0 & 0 & 0 \\ \tilde{x}_{2,1} & 0 & 0 & 0 \\ 0 & 0 & 0 & \tilde{x}_{2,2} \\ 0 & 0 & 0 & 0 \end{pmatrix}$$

where the $x_{j,k}$ and $\tilde{x}_{j,k}$ do not vanish.

Then

$$(5.127)\quad w = v(x_{2,1})^* v(x_{1,1}) = -\, v(\tilde{x}_{1,1})^* v(\tilde{x}_{2,1})$$

$$(5.128)\quad u = v(\tilde{x}_{1,1}) v(x_{2,1})^* = -\, v(\tilde{x}_{2,1}) v(x_{1,1})^*$$

(5.129) $\sigma = v(\tilde{x}_{1,2})^* v(x_{2,2}) = -v(\tilde{x}_{2,2})^* v(x_{1,2})$

(5.130) $\tau = v(x_{2,2})\, v(x_{1,2})^* = -\, v(\tilde{x}_{1,2})\, v(x_{2,2})^*$

and

(5.131) $x_{2,1} = x_{1,1}w^*$, $\tilde{x}_{2,1} = -\, u\, x_{1,1}$, $\tilde{x}_{1,1} = u\, x_{1,1}w^*$;

(5.132) $x_{2,2} = \tau x_{1,2}$, $\tilde{x}_{2,2} = -\, x_{1,2}\sigma^*$, $\tilde{x}_{1,2} = \tau x_{1,2}\sigma^*$.

Proof: (i) We prove only that if $(x_1,x_2;\tilde{x}_1,\tilde{x}_2)$ satisfies M then there exist $0 \leq \lambda,\mu \leq 1$ with $\lambda + \mu = 1$ so that (*), ($\overset{*}{*}$) and ($\overset{**}{*}$) hold. The other direction of part (i) of the proposition is trivial.

Now, $x_k \perp \tilde{x}_k$ for $k = 1,2$, and $x_1 \mathrel{G} x_2$, $x_1 \mathrel{G} \tilde{x}_2$, $\tilde{x}_1 \mathrel{G} x_2$ and $\tilde{x}_1 \mathrel{G} \tilde{x}_2$. Since $\operatorname{span}\{b_k,\tilde{b}_k\}_{k=1}^2 \perp \operatorname{span}\{a_k,\tilde{a}_k\}_{k=1}^2$, we get that the same relations ho the elements in each of the systems $(a_1,a_2;\tilde{a}_1,\tilde{a}_2)$ and $(b_1,b_2;\tilde{b}_1,\tilde{b}_2)$, and that in each system either all the elements vanish, or none of them vanish. This reduces the proof of (i) in this case to the proof of (ii).

(ii) Let a,b,u,w,σ and τ as in Definition 4.4 of the matrix condition, and assume that we have (4.21), (4.22), (5.125) and (5.126). Now, $x_1 G x_2$, so $\ell(x_{1,1}) = \ell(x_{2,1})$ and $r(x_{1,2}) = r(x_{2,2})$. So $a = \ell(x_2)x_1(1 - r(x_2)) =$ $= x_{1,1}$. Similarly, $b = x_{1,2}$. By symmetry, we get in the same way that $x_{2,1} = aw^*$, $x_{2,2} = \tau b$, $\tilde{x}_{2,1} = -ua$, $\tilde{x}_{2,2} = -\, b\sigma^*$, $\tilde{x}_{1,1} = uaw^*$ and $\tilde{x}_{1,2} = \tau b\sigma^*$. So, we have (5.131) and (5.132). From this it is clear that (5.127) - (5.130) hold. □

We are ready to begin our investigation in case 3 of Proposition 5.3. Recall that in this case the range $R(P)$ of the contractive projection P in C_1 contains a normalized system $(x,y;\tilde{x},\tilde{y})$ which satisfies M, and $G(x+\tilde{x})P = 0$.

Lemma 5.23: Let P be a contractive projection in C_1 so that $X = R(P)$ is irreducible, let x be an atom of P and assume that case 3 of

Proposition 5.3 occurs. Then there exist finite sequences of atoms of P, $\{z_k\}_{k=1}^n$ and $\{\tilde{z}_k\}_{k=1}^n$ with $z_1 = x$, so that:

(i) the system $(z_1,\ldots,z_n;\ \tilde{z}_1,\ldots,\tilde{z}_n)$ satisfies SM;

(ii) any sequence $\{y_k\}_{k=1}^n$, such that for every $1 \leq k \leq n$ $y_k \in \{z_k,\tilde{z}_k\}$, satisfies SG;

(iii) put

$$(5.133)\qquad Q_n = \left(\prod_{k=1}^n G(z_k)\, G(\tilde{z}_k)\right) P\ ,$$

then $\dim R(Q_n) \leq 1$;

(iv) we have $\ell(X) = \ell(z_1 + \tilde{z}_1)$ and $r(X) = r(z_1+\tilde{z}_1)$.

If $Q_n = 0$, then $X = \text{span}\{z_k,\tilde{z}_k\}_{k=1}^n$ and for every $c \in C_1$ with $F(X)c = 0$, we have

$$(5.134)\qquad Pc = \sum_{k=1}^n (\langle c,v(z_k)\rangle\, z_k + \langle c,v(\tilde{z}_k)\rangle\, \tilde{z}_k)$$

If $\dim R(Q_n) = 1$, there exists an atom y of Q_n so that $X = \text{span}\{z_1,\ldots,z_n;\tilde{z}_1,\ldots,\tilde{z}_n;y\}$, the system $(z_1,\ldots,z_n;\tilde{z}_1,\ldots,\tilde{z}_n;y)$ satisfies DSM, and for every $c \in C_1$ with $F(X)c = 0$ we have

$$(5.135)\qquad Pc = \sum_{k=1}^n (\langle c,v(z_k)\rangle\, z_k + \langle c,v(\tilde{z}_k)\rangle\, \tilde{z}_k) + \langle c,v(y)\rangle y$$

<u>Proof:</u> Let $x,y,\tilde{x},\tilde{y}$ be atoms of P so that the system $(x,y;\tilde{x},\tilde{y})$ satisfies M, and so that $G(x+\tilde{x})P = 0$.

We construct the sequences $\{z_k\}_{k=1}^n$ and $\{\tilde{z}_k\}_{k=1}^n$ of atoms of P by induction so that (i) and (ii) are satisfied, and so that if $E_k,F_k,G_k,\tilde{E}_k,\tilde{F}_k$ and $\tilde{G}_k$ denote respectively $E(z_k)$, $F(z_k)$, $G(z_k)$, $E(\tilde{z}_k)$, $F(\tilde{z}_k)$, and $G(\tilde{z}_k)$, and if Q_n is defined by (5.133), then

$$(5.136)\qquad P = \sum_{j=1}^n (E_jP + \tilde{E}_jP) + Q_n$$

We put $z_1 = x$, $z_2 = y$, $\tilde{z}_1 = \tilde{x}$ and $\tilde{z}_2 = \tilde{y}$. Then (i) and (ii) are satisfied with $n = 2$. Now X is irreducible, so from $G(z_1+\tilde{z}_1)P = 0$ we deduce that $F(z_1+\tilde{z}_1)P = 0$, and that (iv) holds. Moreover, $z_2\ G\ z_1$ and $z_2\ G\ \tilde{z}_1$ imply by the atomicity of z_1 and $\tilde{z}_1$ that $E_2Q_1 = E_2P$, and similarly: $\tilde{E}_2Q_1 = \tilde{E}_2P$.
So,

$$\begin{aligned}(5.137)\qquad P &= E(z_1+\tilde{z}_1)P = E_1P + \tilde{E}_1P + Q_1 \\ &= E_1P + \tilde{E}_1P + \tilde{E}_2P + Q_2\end{aligned}$$

which proves (5.136) for $n = 2$.

Assume that we have constructed atoms $\{z_j\}_{j=1}^k$ and $\{\tilde{z}_j\}_{j=1}^k$ of P so that (i), (ii) and (5.136) are satisfied with k instead of n. If $Q_k = 0$, we put $n = k$, and then for every $c \in C_1$ we get by (5.136) and the atomicity of the z_j and $\tilde{z}_j$:

$$(5.138)\qquad Pc = \sum_{j=1}^n (E_jPc + \tilde{E}_jPc) = \sum_{j=1}^n (\langle c,P^*v(z_j)\rangle\, z_j + \langle c,P^*v(\tilde{z}_j)\rangle\, \tilde{z}_j)$$

So, $X = \text{span}\,\{z_j,\tilde{z}_j\}_{j=1}^n$. Also, by Proposition 5.5 we have for every $1 \leq j \leq n$ that $P^*v(z_j) \perp v(\tilde{z}_j)$ and that $P^*v(\tilde{z}_j) \perp v(z_j)$. Using Propositions 1.5 and 1.6 we get that for every $1 \leq j \leq n$

$$(5.139)\qquad (G(X) + E(X))P^*v(z_j) = v(z_j),\ (G(X) + E(X))P^*\, v(\tilde{z}_j) = v(\tilde{z}_j).$$

If $c \in C_1$ with $F(X)c = 0$, then (5.138) and (5.131) imply (5.134).

If $Q_k \neq 0$, let z be any atom of Q_k. As in Lemma 4.10, either $E(z) = E(z_1 + \tilde{z}_1) = E(X)$, and then $Q_k = E(z)Q_k$ and Q_k is a one-dimensional projection, or Q_k has besides z another atom $\tilde{z}$ so that z and $\tilde{z}$ are atoms of P, $z \perp \tilde{z}$ and for $1 \leq j \leq k$ we have $z\ G\ z_j$, $z\ G\ \tilde{z}_j$, $\tilde{z}\ G\ z_j$ and $\tilde{z}\ G\ \tilde{z}_j$.
In the first case we put $n = k$, and as in the proof of Lemma 4.10 in the first case, we can choose $y = \theta z$, $|\theta| = 1$, so that $(z_1,\tilde{z}_1,y)$ satisfies D.

If $1 < j \leqslant n$, let $\tilde{P} = (\prod_{\substack{i=2 \\ i \neq j}}^{m} G_i \tilde{G}_i)P$. Note that $z_1, z_j, \tilde{z}_1, \tilde{z}_j$ are atoms of $\tilde{P}$ (being atoms of P and elements of $R(\tilde{P})$), that y is an atom of $G_1\tilde{G}_1G_j\tilde{G}_j\tilde{P} = Q_n$, and that the system $(z_1, z_j; \tilde{z}_1, \tilde{z}_j)$ satisfies M. By our choice of $y = \theta z$, we get as in the proof of the first case in Lemma 4.10, that the system $(z_j, \tilde{z}_j, y)$ satisfies D. Since j is arbitrary, we get that $(z_1, \ldots, z_n; \tilde{z}_1, \ldots, \tilde{z}_n; y)$ satisfies DSM. Also, (5.136) and the atomocity of the $\{z_j, \tilde{z}_j\}_{j=1}^{n}$ with respect to P and the atomocity of y with respect to Q_n imply that for every $c \in C_1$:

$$(5.140) \quad Pc = \sum_{j=1}^{n} (E_j Pc + \tilde{E}_j Pc) + E(y)Q_n c$$

$$= \sum_{j=1}^{n} (\langle c, P^* v(z_j)\rangle\, z_j + \langle c, P^* v(\tilde{z}_j)\rangle\, \tilde{z}_j) + \langle c, P^* v(y)\rangle\, y$$

This implies that $X = \operatorname{span}\{z_1, \ldots, z_n; \tilde{z}_1, \ldots, \tilde{z}_n; y\}$. Of course, (5.139) is valid in this case, and $(G(X) + E(X))P^* v(y) = v(y)$ follows by Propositions 1.5 and 1.6 and by the fact that $\ell(X) = \ell(y)$ and $r(X) = r(y)$. Thus for every $c \in C_1$ with $F(X)c = 0$ we have (5.135) and by (5.139), (5.140) and by $(G(X)+E(X))P^* v(y) = v(y)$.

In case $\dim R(Q_k) \geqslant 2$, let $1 < j \leqslant k$ and put $\tilde{P} = (\prod_{\substack{i=2 \\ i \neq j}}^{k} G_i \tilde{G}_i)P$. Again, $(z_1, z_j, \tilde{z}_1, \tilde{z}_j)$ satisfies M, and z and $\tilde{z}$ are atoms of $G_1\tilde{G}_1G_j\tilde{G}_j\tilde{P} = Q_k$. As in the proof of the second case of Lemma 4.10 we can choose z and $\tilde{z}$ so that $(z_1, z; \tilde{z}_1, \tilde{z})$ satisfies M, and this gives us automatically that also $(z_j, z; \tilde{z}_j, \tilde{z})$ satisfies M. Put $z_{k+1} = z$, $\tilde{z}_{k+1} = \tilde{z}$. Since the index j was arbitrary, we get that the system $(z_1, \ldots, z_{k+1}; \tilde{z}_1, \ldots, \tilde{z}_{k+1})$ satisfies SM.. Thus we have (i) with $n = k+1$. To prove (ii) for $n = k+1$, let $\{y_j\}_{j=1}^{k+1}$ be any sequence satisfying $y_j \in \{z_j, \tilde{z}_j\}$, $1 \leqslant j \leqslant k+1$. By the induction hypothesis $\{y_j\}_{j=1}^{k}$ satisfies SG. Also, $y_j \; G \; y_{k+1}$ for $1 \leqslant j \leqslant k$. Thus, as in the proof of Lemma 5.11 we get that $\{y_j\}_{j=1}^{k+1}$ satisfies SG.

In order to prove (5.136) for $n = k+1$ it is clearly enough to show that

$$Q_k = E_{k+1}P + \tilde{E}_{k+1}P + Q_{k+1}, \tag{5.141}$$

and this follows from $E_{k+1}Q_k = E_{k+1}P$ and $\tilde{E}_{k+1}Q_k = E_kP$ by the same arguments which yield this for $k = 1$.

So (i), (ii) and (5.136) are satisfied also for $n = k+1$.

In order to complete the proof of the lemma it is thus enough to prove that for some $n < \infty$ we have $\dim (Q_n) \leq 1$. If not, we get infinite sequences $\{z_j\}_{j=1}^\infty$ and $\{\tilde{z}_j\}_{j=1}^\infty$ of normalized elements which satisfy in particular SG. By the analysis in subsection 5(e) we get (5.116) for every $1 \leq k < \infty$. In particular, we get $F_1(\prod_{j=2}^\infty G_j) = 0$, and this leads to a contradiction since $\tilde{z}_1 = F_1(\prod_{j=2}^{\tilde{\ }} G_j)z_1$.

This completes the proof of the lemma. □

Let $\{z_k\}_{k=1}^n$ and $\{\tilde{z}_k\}_{k=1}^n$, $2 \leq n < \infty$, be as in Lemma 5.23. We shall show that if $Q_n = 0$ then X is an elementary subspace of C_1 of type 5, i.e. $X = AH_1^n(a,\tilde{a})$ for some $a,\tilde{a} \in C_1$, and that if $\dim R(Q_n) = 1$, then X is an elementary subspace of C_1 of type 6, i.e. $X = DAH_1^{n+1}(a)$ for some $a \in C_1$. We consider the first case before.

We continue to denote $I_n = \{1,2,\ldots,n\}$, $\mathcal{P}_n$ is the set of all subsets of I_n, and $\mathcal{P}_n^k = \{\alpha \in \mathcal{P}_n,\ k \in \alpha\}$, $1 \leq k \leq n$. For every $\alpha \in \mathcal{P}_n$, put

$$\ell_\alpha = \prod_{\substack{k\in\alpha \\ j\in I_n\sim\alpha}} (\ell(z_k)\ \ell(\tilde{z}_j)\ (1 - \ell(z_j))\ (1 - \ell(\tilde{z}_k))) \tag{5.142}$$

$$r_\alpha = \prod_{\substack{k\in\alpha \\ j\in I_n\sim\alpha}} r(\tilde{z}_k)\ r(z_j)\ (1-r(\tilde{z}_j))\ (1-r(z_k)). \tag{5.143}$$

If $\alpha \in \mathcal{P}_n^k$ define

$$z_{k,\alpha} = \ell_\alpha \cdot z_k \cdot r_{\alpha\sim\{k\}};\ \tilde{z}_{k,\alpha} = \ell_{\alpha\sim\{k\}} \cdot \tilde{z}_k \cdot r_\alpha \tag{5.144}$$

We state without proof the following proposition. The proof is straightforward and follows directly from the definitions.

Proposition 5.24: Let $\{z_k\}_{k=1}^n$ and $\{\tilde{z}_k\}_{k=1}^n$ be normalized elements in C_1 so that the system $(z_1,\dots,z_n;\tilde{z}_1,\dots,\tilde{z}_n)$ satisfies SM. Then:

(i) $\ell(z_{k,\alpha}) = \ell_\alpha$ and $r(\tilde{z}_{k,\alpha}) = r_\alpha$, $\alpha \in \mathcal{P}_n^k$;

(ii) $r(z_{k,\alpha}) = r_{\alpha\sim\{k\}}$ and $\ell(\tilde{z}_{k,\alpha}) = \ell_{\alpha\sim\{k\}}$, $\alpha \in \mathcal{P}_n^k$;

(iii) if $\alpha,\beta \in \mathcal{P}_n$ and $\alpha \neq \beta$, then $\ell_\alpha \cdot \ell_\beta = 0 = r_\alpha \cdot r_\beta$;

(iv) if $\alpha,\beta \in \mathcal{P}_n^k$ and $\alpha \neq \beta$ then $z_{k,\alpha} \perp z_{k,\beta}$ and $\tilde{z}_{k,\alpha} \perp \tilde{z}_{k,\beta}$;

(v) $z_k = \sum_{\alpha\in\mathcal{P}_n^k} z_{k,\alpha}$; $\tilde{z}_k = \sum_{\alpha\in\mathcal{P}_n^k} \tilde{z}_{k,\alpha}$.

(vi) $1 = \sum_{\alpha\in\mathcal{P}_n^k} \|z_{k,\alpha}\|_1 = \sum_{\alpha\in\mathcal{P}_n^k} \|\tilde{z}_{k,\alpha}\|_1$

(vii) If $\alpha \in \mathcal{P}_n^k$, $\beta \in \mathcal{P}_n^j$ and if $|\alpha| \equiv |\beta| \pmod 2$, then $z_{k,\alpha} \perp \tilde{z}_{j,\beta}$; if $|\alpha| \equiv |\beta|+1 \pmod 2$, then $z_{k,\alpha} \perp z_{j,\beta}$ and $\tilde{z}_{k,\alpha} \perp \tilde{z}_{j,\beta}$.

In what follows we shall use only the fact that the system $(z_1,\dots,z_n;\tilde{z}_1,\dots,\tilde{z}_n)$ satisfies SM in order to prove the existence of elements $a,\tilde{a} \in C_1$ with $a \perp \tilde{a}$ and $\|a\|_1 + \|\tilde{a}\|_1 = 2^{2-n}$, and the existence of a tensor product representation in which for every $1 \leq k \leq n$:

(5.145) $z_k = x_{n,k} \otimes a + (\tilde{x}_{n,k})^T \otimes \tilde{a}$;

(5.146) $\tilde{z}_k = \tilde{x}_{n,k} \otimes a + x_{n,k}^T \otimes \tilde{a}$;

where the matrices $x_{n,k}$ and $\tilde{x}_{n,k}$ were defined by (2.38) and (2.39) respectively and "T" denotes transpose.

We put

(5.147) $\mathcal{P}_{n,(0)} = \{\alpha \in \mathcal{P}_n; \ |\alpha| \text{ is an odd number}\}$

(5.148) $\mathcal{P}_{n,(e)} = \{\alpha \in \mathcal{P}_n; \ |\alpha| \text{ is an even number}\}$

(5.149) $\mathcal{P}^k_{n,(0)} = \mathcal{P}_{n,(0)} \cap \mathcal{P}^k_n; \ \mathcal{P}^k_{n,(e)} = \mathcal{P}_{n,(e)} \cap \mathcal{P}^k_n$.

Formulas (5.145) and (5.146) will follow from the following lemma:

Lemma 5.25: Let $\{z_k\}^n_{k=1}$ and $\{\tilde{z}_k\}^n_{k=1}$ be normalized elements in C_1 so that $(z_1,\ldots,z_n;\tilde{z}_1,\ldots,\tilde{z}_n)$ satisfies SM. Put $a = z_{1,\{1\}}$ and $\tilde{a} = \tilde{z}_{1,\{1\}}$. Then there exist partial isometries $\{u_\alpha\}_{\alpha\in\mathcal{P}_{n,(0)}}$, $\{\tau_\alpha\}_{\alpha\in\mathcal{P}_{n,(e)}}$, $\{w_\alpha\}_{\alpha\in\mathcal{P}_{n,(e)}}$ and $\{\sigma_\alpha\}_{\alpha\in\mathcal{P}_{n,(0)}}$, so that for any $\alpha \in\mathcal{P}_{n(0)}$:

(5.150) $r(u_\alpha) = \ell_{\{1\}} = u_{\{1\}} \ ; \ \ \ell(u_\alpha) = \ell_\alpha \ ;$

(5.151) $r(\sigma_\alpha) = r_{\{1\}} = \sigma_{\{1\}} \ ; \ \ \ell(\sigma_\alpha) = r_\alpha \ ;$

and for any $\alpha \in\mathcal{P}_{n,(e)}$:

(5.152) $r(\tau_\alpha) = \ell_\phi = \tau_\phi \ ; \ \ \ell(\tau_\alpha) = \ell_\alpha \ ;$

(5.153) $r(w_\alpha) = r_\phi = w_\phi \ ; \ \ \ell(w_\alpha) = r_\alpha \ ;$

and so that for every $1 \leq k \leq n$ and $\alpha \in\mathcal{P}^k_{n,(0)}$:

(5.154) $z_{k,\alpha} = (-1)^{i(\alpha,k)} \cdot u_\alpha \, a \, w^*_{\alpha\sim\{k\}}$

(5.155) $\tilde{z}_{k,\alpha} = (-1)^{i(\alpha,k)} \cdot \tau_{\alpha\sim\{k\}} \cdot \tilde{a} \, \sigma^*_\alpha$

and for every $1 \leq k \leq n$ and $\alpha \in \mathcal{P}^k_{n,(e)}$:

$$(5.156) \qquad z_{k,\alpha} = (-1)^{i(\alpha,k)} \cdot \tau_\alpha \tilde{a} \sigma^*_{\alpha \sim \{k\}} ;$$

$$(5.157) \qquad \tilde{z}_{k,\alpha} = (-1)^{i(\alpha,k)} \cdot u_{\alpha \sim \{k\}} \cdot a w^*_\alpha .$$

The following matrix diagrams describe the situation in Lemma 5.25, and can help the reader to follow the proof. Here $\alpha \in \mathcal{P}^k_{n,(e)}$ and $\beta \in \mathcal{P}^k_{n,(0)}$.

$$(5.158) \qquad z_{k,\alpha} + z_{k,\beta} =$$

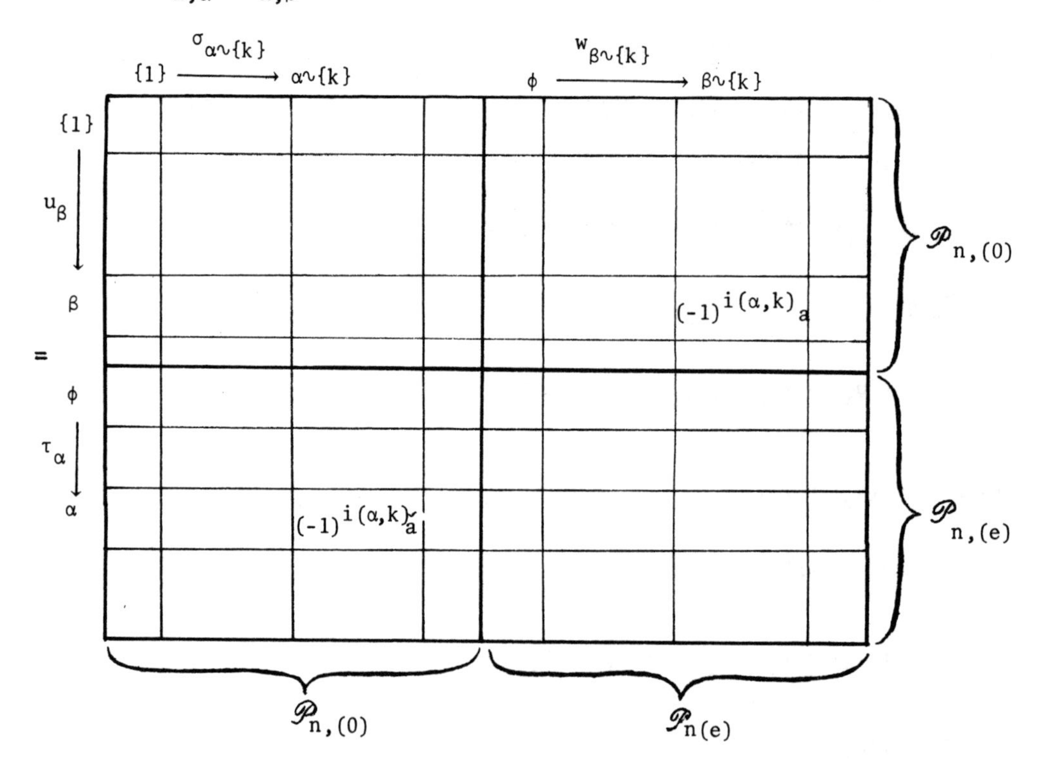

(5.159) $\quad \tilde{z}_{k,\alpha} + \tilde{z}_{k,\beta} =$

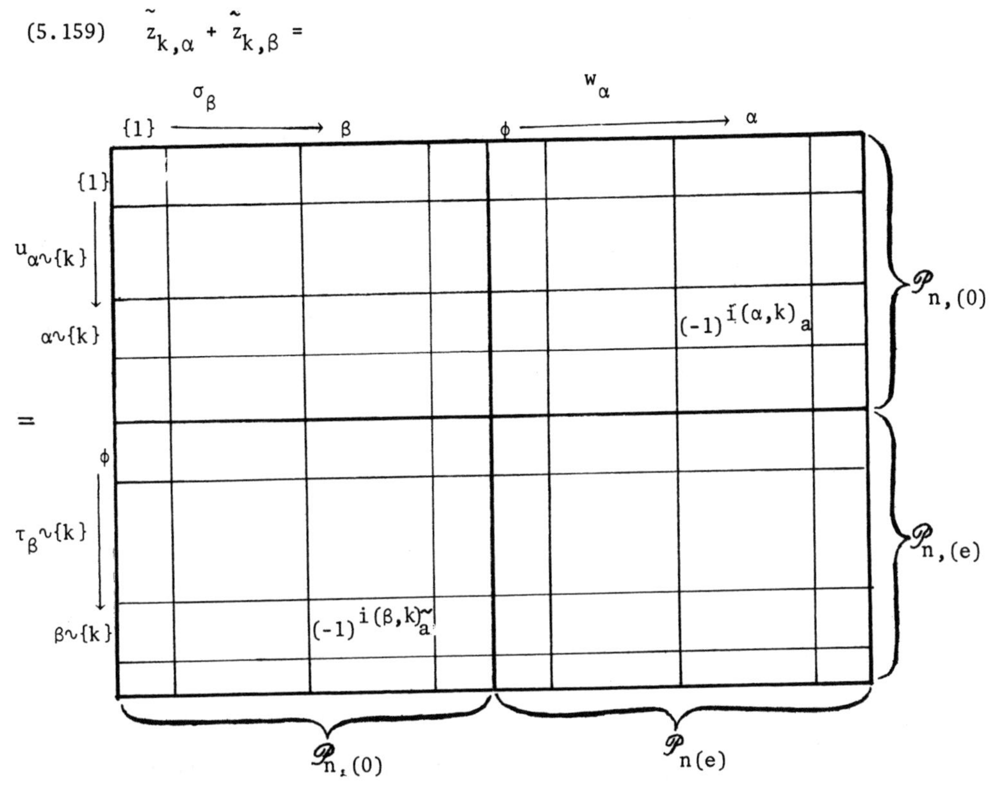

<u>Proof of Lemma 5.25:</u> We prove the lemma by induction on n. For $n = 2$ the assertions of the lemma follow directly from the definition of the matrix condition. Assume that $3 \leq n$ and that the statement of the lemma holds for $n-1$, and let $(z_1,\ldots,z_n;\tilde{z}_1,\ldots,\tilde{z}_n)$ satisfies SM (in particular, $\|z_k\|_1 = 1 = \|\tilde{z}_k\|_1$).

Put for $1 \leq k \leq n-1$

(5.160) $\quad a_k = \ell(\tilde{z}_n) z_k \, r(z_n) \;;\quad \tilde{a}_k = \ell(\tilde{z}_n)\tilde{z}_k \, r(z_n) \;;$

(5.161) $\quad b_k = \ell(z_n) \, z_k \, r(\tilde{z}_n) \;;\quad \tilde{b}_k = \ell(z_n) \, \tilde{z}_k \, r(\tilde{z}_n).$

Since for $1 \leq k \leq n-1$ the system $(z_k,z_n;\, \tilde{z}_k,\tilde{z}_n)$ satisfies M, we clearly have

(5.162) $\quad z_k = a_k + b_k \;;\quad \tilde{z}_k = \tilde{a}_k + \tilde{b}_k$

and for any $1 \leq j,k \leq n-1$:

(5.163) $a_k \perp b_j, \quad a_k \perp \tilde{b}_j, \quad \tilde{a}_k \perp b_j, \quad \tilde{a}_k \perp \tilde{b}_j.$

Also, by Proposition 5.22 and the fact that the system $(z_1,\ldots,z_{n-1};\tilde{z}_1,\ldots,\tilde{z}_{n-1})$ satisfies SM, we get the existence of $0 \leq \lambda,\mu \leq 1$ with $\lambda + \mu = 1$, so that

(5.164) $\|a_k\|_1 = \lambda = \|\tilde{a}_k\|_1 \; ; \; \|b_k\|_1 = \mu = \|\tilde{b}_k\|_1$

(5.165) if $\lambda \neq 0$, then $(\lambda^{-1}a_1,\ldots,\lambda^{-1}a_{n-1};\ \lambda^{-1}\tilde{a}_1,\ldots,\lambda^{-1}\tilde{a}_{n-1})$

satisfies SM;

(5.166) if $\mu \neq 0$, then $(\mu^{-1}b_1,\ldots,\mu^{-1}b_{n-1};\mu^{-1}\tilde{b}_1,\ldots,\mu^{-1}\tilde{b}_{n-1})$

satisfies SM.

If $\lambda\cdot\mu = 0$, then we get $E(z_1+\tilde{z}_1) < E(z_n+\tilde{z}_n)$, which contradicts the fact that $(z_1,z_n;\tilde{z}_1,\tilde{z}_n)$ satisfies M. So $0 < \lambda,\mu < 1$.

Put $a = z_{1,\{1\}}$, $\tilde{a} = \tilde{z}_{1,\{1\}}$. We apply the induction hypothesis on the system $(\lambda^{-1}a_1,\ldots,\lambda^{-1}a_{n-1};\ \lambda^{-1}\tilde{a}_1,\ldots,\lambda^{-1}\tilde{a}_{n-1})$ and get partial isometries:

(5.167) $\{u_\alpha\}_{\alpha\in\mathcal{P}_{n-1,(0)}}, \quad \{\tau_\alpha\}_{\alpha\in\mathcal{P}_{n-1,(e)}}, \quad \{w_\alpha\}_{\alpha\in\mathcal{P}_{n-1,(e)}}, \quad \{\sigma_\alpha\}_{\alpha\in\mathcal{P}_{n-1,(0)}}$

so that (5.150) - (5.157) are satisfied with $n-1$ instead of n (note that for $\alpha \in \mathcal{P}^k_{n-1}$ we have $a_{k,\alpha} = z_{k,\alpha}$ and $\tilde{a}_{k,\alpha} = \tilde{z}_{k,\alpha}$).

Now $(z_1,z_n;\tilde{z}_1,\tilde{z}_n)$ satisfies M, so:

(5.168) $u = v(z_{n,\{n\}})v(a)^* = -\ v(\tilde{z}_{1,\{1,n\}})\ v(\tilde{z}_{n,\{1,n\}})^* \;;$

(5.169) $\tau = v(z_{1,\{1,n\}}) \cdot v(\tilde{z}_{n,\{n\}})^* = -v(z_{n,\{1,n\}})v(\tilde{a})^* \;;$

(5.170) $w = v(\tilde{z}_{1,\{1,n\}})^*\ v(z_{n,\{n\}}) = -\ v(\tilde{z}_{n,\{1,n\}})^*\ v(a);$

(5.171) $\sigma = v(\tilde{z}_{n,\{n\}})\ v(\tilde{a}) = -\ v(z_{1,\{1,n\}})^*\ v(z_{n,\{1,n\}}).$

(5.172) $\quad z_{n,\{n\}} = ua, \; \tilde{z}_{n,\{1,n\}} = - aw^*$

(5.173) $\quad \tilde{z}_{1,\{1,n\}} = \tilde{b}_{1,\{1,n\}} = u\, a\, w^*$;

(5.174) $\quad z_{n,\{1,n\}} = - \tau\tilde{a} \; , \; \tilde{z}_{n,\{n\}} = \tilde{a}\sigma$;

(5.175) $\quad z_{1,\{1,n\}} = b_{1,\{1,n\}} = \tau\, \tilde{a}\, \sigma^*$.

Note that the mapping $\alpha \to \alpha \cup \{n\}$ is a one-to-one mapping of $\mathcal{P}_{n-1,(0)}$ onto $\mathcal{P}^{(n)}_{n,(e)}$ and also a one-to-one mapping of $\mathcal{P}_{n-1,(0)}$ onto $\mathcal{P}^{n}_{n,(e)}$. Thus, by the induction hypothesis and the fact that the system $(\mu^{-1}b_1,\ldots,\mu^{-1}b_{n-1};\mu^{-1}\tilde{b}_1,\ldots,\mu^{-1}\tilde{b}_{n-1})$ satisfies SM we get partial isometries

(5.176) $\quad \{\tilde{u}_\alpha\}_{\alpha\in\mathcal{P}^{n}_{n,(0)}} \; , \; \{\tilde{\tau}_\alpha\}_{\alpha\in\mathcal{P}^{n}_{n,(e)}} \; , \; \{\tilde{w}_\alpha\}_{\alpha\in\mathcal{P}^{n}_{n,(e)}} \; , \; \{\tilde{\sigma}_\alpha\}_{\alpha\in\mathcal{P}^{n}_{n,(0)}}$

so that if we put $b = z_{1,\{1,n\}}$ and $\tilde{b} = \tilde{z}_{1,\{1,n\}}$ then for every $\alpha \in\mathcal{P}^{n}_{n,(0)}$:

(5.177) $\quad r(\tilde{u}_\alpha) = \ell(\tilde{b}) = \tilde{u}_{\{n\}}$; $\ell(\tilde{u}_\alpha) = \ell_\alpha$;

(5.178) $\quad r(\tilde{\sigma}_\alpha) = r(b) = \tilde{\sigma}_{\{n\}}$; $\ell(\tilde{\sigma}_\alpha) = r_\alpha$;

and for every $\alpha \in\mathcal{P}^{n}_{n,(e)}$:

(5.179) $\quad r(\tilde{\tau}_\alpha) = \ell(b) = \tilde{\tau}_{\{1,n\}}$; $\ell(\tilde{\tau}_\alpha) = \ell_\alpha$;

(5.180) $\quad r(\tilde{w}_\alpha) = r(\tilde{b}) = \tilde{w}_{\{1,n\}}$; $\ell(\tilde{w}_\alpha) = r_\alpha$;

and so that for every $1 \leq k \leq n-1$ and every $\alpha \in\mathcal{P}^{k}_{n,(0)} \cap \mathcal{P}^{n}_{n,(0)}$:

(5.181) $\quad z_{k,\alpha} = (-1)^{i(\alpha,k)} \cdot \tilde{u}_\alpha\, \tilde{b}\, \tilde{w}_{\alpha\sim\{k\}}$;

(5.182) $\tilde{z}_{k,\alpha} = (-1)^{i(\alpha,k)} \cdot \tilde{\tau}_{\alpha\sim\{k\}}\, b\, \tilde{\sigma}_{\alpha}^{*}$;

and for every $1 \leq k \leq n-1$ and every $\alpha \in \mathcal{P}_{n,(e)}^{k} \cap \mathcal{P}_{n,(e)}^{n}$:

(5.183) $z_{k,\alpha} = (-1)^{i(\alpha,k)} \cdot \tilde{\tau}_{\alpha}\, b\, \tilde{\sigma}_{\alpha\sim\{k\}}^{*}$;

(5.184) $\tilde{z}_{k,\alpha} = (-1)^{i(\alpha,k)} \cdot \tilde{u}_{\alpha\sim\{k\}}\, \tilde{b}\, \tilde{w}_{\alpha}^{*}$;

Put for $\alpha \in \mathcal{P}_{n,(0)}^{n}$:

(5.185) $u_{\alpha} = \tilde{u}_{\alpha} u$; $\sigma_{\alpha} = \tilde{\sigma}_{\alpha}\, \sigma$;

and for $\alpha \in \mathcal{P}_{n,(e)}^{n}$:

(5.186) $w_{\alpha} = \tilde{w}_{\alpha}\, w$; $\tau_{\alpha} = \tilde{\tau}_{\alpha}\, \tau$.

Then, clearly, the families

(5.187) $\{u_{\alpha}\}_{\alpha\in\mathcal{P}_{n,(0)}}$, $\{\sigma_{\alpha}\}_{\alpha\in\mathcal{P}_{n,(0)}}$, $\{w_{\alpha}\}_{\alpha\in\mathcal{P}_{n,(e)}}$, $\{\tau_{\alpha}\}_{\alpha\in\mathcal{P}_{n,(e)}}$

satisfy (5.150) - (5.153), and formulas (5.154) - (5.157) are satisfied for $1 \leq k \leq n-1$ and any $\alpha \in \mathcal{P}_{n}^{k}$, and for $k = n$ and $\alpha = \{n\}$ or $\alpha = \{1,n\}$. Call $\alpha \in \mathcal{P}_{n}^{n}$ "good for z_n" (respectively, "good for $\tilde{z}_n$") if (5.154) or (5.156) (respectively, if (5.155) or (5.157)) is satisfied with α and $k = n$. Call $\alpha \in \mathcal{P}_{n}^{n}$ "good" if it is "good for z_n" and "good for $\tilde{z}_n$". Since $\{n\}$ and $\{1,n\}$ are "good", in order to show that every $\alpha \in \mathcal{P}_{n}^{n}$ is "good" it is clearly enough to prove the following three assertions:

(i) if $\alpha \in \mathcal{P}_{n}^{n}$ is "good", then every $\beta \in \mathcal{P}_{n}^{n}$ with $|\alpha| = |\beta|$ is also "good";

(ii) if $\alpha \in \mathcal{P}_{n}^{n}$ is "good for z_n", and if $k \notin \alpha$, then $\alpha \cup \{k\}$ is "good for $\tilde{z}_n$" ;

(iii) if $\alpha \in \mathscr{P}_n^n$ is "good for $\tilde{z}_n$", and if $k \notin \alpha$ then $\alpha \cup \{k\}$ is "good for z_n".

Since $\{z_k\}_{k=1}^n$ and $\{\tilde{z}_k\}_{k=1}^n$ satisfy SG, for every $1 \leq m \leq n$ the sequences $\{z_{k,m}\}_{k=1}^n$ and $\{\tilde{z}_{k,m}\}_{k=1}^m$ satisfy SG_m (see formula (5.73), Proposition 5.14 and Definition 5.15). Thus assertion (i) follows by the first step in the proof of Lemma 5.17.

Let us prove (ii). Assume that $\alpha \in \mathscr{P}_n^n$ is "good for z_n", and that $k \notin \alpha$. Put $\beta = \alpha \cup \{k\}$, $\alpha' = (\alpha \sim \{n\}) \cup \{k\}$. Since the system $(z_k, z_n; \tilde{z}_k, \tilde{z}_n)$ satisfies M, we get that $\|z_{k,\alpha'}\|_1 = \|z_{n,\alpha}\|_1 = \|z_{k,\beta}\|_1 = \|\tilde{z}_{n,\beta}\|_1 = t$, $t > 0$, and that the system $(t^{-1}z_{k,\alpha'}, t^{-1}z_{n,\alpha}; t^{-1}z_{k,\beta}, t^{-1}\tilde{z}_{n,\beta})$ satisfies M. This implies in particular that $|\tilde{z}_{n,\beta}| = |\tilde{z}_{k,\beta}|$ and that

$$(5.188) \qquad v(\tilde{z}_{n,\beta})\, v(\tilde{z}_{k,\beta})^* = -\, v(z_{k,\alpha'})\, v(z_{n,\alpha})^*$$

Assume for example that $\alpha \in \mathscr{P}_{n,(0)}^n$, the proof in the case $\alpha \in \mathscr{P}_{n,(e)}^n$ is the same and differs only in the notations. Now from (5.154) for n and α and for k and α' we get

$$(5.189) \qquad v(z_{n,\alpha}) = (-1)^{i(\alpha,n)} \cdot u_\alpha\, v(a)\, w^*_{\alpha\sim\{n\}}$$

$$(5.190) \qquad v(z_{k,\alpha'}) = (-1)^{i(\alpha,k)}\, u_{\alpha'}\, v(a)\, w^*_{\alpha\sim\{n\}}$$

So:

$$(5.191) \qquad \tilde{z}_{n,\beta} = v(\tilde{z}_{n,\beta})\, v(\tilde{z}_{k,\beta})^*\, \tilde{z}_{k,\beta}$$

$$= -v(z_{k,\alpha'})\, v(z_{n,\alpha})^* (-1)^{i(\beta,k)} \cdot u_\alpha\, a\, w^*_\beta$$

$$= -(-1)^{i(\alpha,n)} (u_{\alpha'}\, v(a)\, w^*_{\alpha\sim\{n\}})\, (w_{\alpha\sim\{n\}} v(a^*) u^*_\alpha)\, (u_\alpha\, a\, w^*_\beta)$$

$$=-(-1)^{i(\alpha,n)} u_{\alpha}, a w_{\beta}^{*} = (-1)^{i(\beta,n)} \cdot u_{\beta\sim\{n\}} a w_{\beta}^{*}$$

This completes the proof of (ii).

Now, (iii) follows from (ii) by the symmetry between the $\{z_k\}_{k=1}^n$ and the $\{\tilde{z}_k\}_{k=1}^n$, and so the proof of Lemma 5.25 is complete. □

Let us sum the results obtained so far in this subsection in the following:

Corollary 5.26: (i) Let $\{z_k\}_{k=1}^n$, $\{\tilde{z}_k\}_{k=1}^n$ be normalized elements in C_1 so that the system $(z_1,\ldots,z_n;\tilde{z}_1,\ldots,\tilde{z}_n)$ satisfies SM. Then there exists a tensor product representation and elements $a,\tilde{a} \in C_1$ with $a \perp \tilde{a}$ and $\|a\|_1 + \|\tilde{a}\|_1 = 2^{2-n}$ so that:

$$(5.192) \quad z_k = x_{n,k} \otimes a + (\tilde{x}_{n,k})^T \otimes \tilde{a}$$

$$(5.193) \quad \tilde{z}_k = \tilde{x}_{n,k} \otimes a + x_{n,k}^T \otimes \tilde{a}$$

where the $x_{n,k}$ and the $\tilde{x}_{n,k}$ were defined by (2.38) and (2.39) respectively. In particular, there exists a contractive projection from C_1 onto $\mathrm{span}(\{z_k\}_{k=1}^n \cup \{\tilde{z}_k\}_{k=1}^n)$, which is canonical in the sense of §2c.

(ii) Let P be a contractive projection in C_1 so that $X = R(P)$ is irreducible. Assume that case 3 of Proposition 5.3 occurs and the case "$Q_n = 0$" of Lemma 5.23 occurs. Then $X = \mathrm{span}(\{z_k\}_{k=1}^n \cup \{\tilde{z}_k\}_{k=1}^n)$, where the system $(z_1,\ldots,z_n;\tilde{z}_1,\ldots,\tilde{z}_n)$ satisfies SM, $PG(X) = 0$, and $PE(X)$ is the canonical contractive projection from C_1 onto X. So, X is an elementary subspace of C_1 of type 5, i.e. space of the form $AH_1^n(a,\tilde{a})$.

Proof: (i) From Lemma 5.25 we get $a,\tilde{a} \in C_1$ with $\|a\|_1 + \|\tilde{a}\|_1 = 2^{2-n}$ and the existence of a tensor product representation in which: for every $\alpha \in \mathcal{P}_{n,(0)}^k$ we have

$$(5.194) \quad z_{k,\alpha} = (-1)^{i(\alpha,k)} \cdot e_{\alpha,\alpha\sim\{k\}} \otimes a$$

(5.195) $$\tilde{z}_{k,\alpha} = (-1)^{i(\alpha,k)} e_{\alpha\sim\{k\},\alpha} \otimes \tilde{a}$$

and for $\alpha \in \mathcal{P}^k_{n,(e)}$:

(5.196) $$z_{k,\alpha} = (-1)^{i(\alpha,k)} \cdot e_{\alpha,\alpha\sim\{k\}} \otimes \tilde{a}$$

(5.197) $$\tilde{z}_{k,\alpha} = (-1)^{i(\alpha,k)} \cdot e_{\alpha\sim\{k\},\alpha} \otimes a$$

So, indeed

(5.198) $$z_k = \sum_{\alpha\in\mathcal{P}^k_{n,(0)}} z_{k,\alpha} + \sum_{\alpha\in\mathcal{P}^k_{n,(e)}} z_{k,\alpha}$$

$$= \Big(\sum_{\alpha\in\mathcal{P}^k_{n,(0)}} (-1)^{i(\alpha,k)} \cdot e_{\alpha,\alpha\sim\{k\}}\Big) \otimes a$$

$$+ \Big(\sum_{\alpha\in\mathcal{P}^k_{n,(e)}} (-1)^{i(\alpha,k)} e_{\alpha,\alpha\sim\{k\}}\Big) \otimes \tilde{a}$$

$$= x_{n,k} \otimes a + (\tilde{x}_{n,k})^T \otimes \tilde{a} .$$

Similarly, (5.193) follows from (5.195) and (5.197). The existence of a canonical contractive projection from C_1 onto $\mathrm{span}(\{z_k\}_{k=1}^n \cup \{\tilde{z}_k\}_{k=1}^n) = AH_1^n(a,\tilde{a})$ was proved in §2c.

(ii) This follows directly from Lemma 5.23 and from (i) of the present corollary. □

Let us continue our investigation with the case "dim $R(Q_n) = 1$" of Lemma 5.23, and let $\{z_k\}_{k=1}^n, \{\tilde{z}_k\}_{k=1}^n$ and y be normalized elements of C_1 so that the system $(z_1,\ldots,z_n;\tilde{z}_1,\ldots,\tilde{z}_n;y)$ satisfies DSM. We shall show that from this condition alone follows the existence of an element $a \in C_1$ with $\|a\|_1 = 2^{1-n}$ and the existence of a tensor-product representation in which

(5.199) $$z_k = x_{n+1,k} \otimes a \ ; \quad \tilde{z}_k = \tilde{x}_{n+1,k} \otimes a$$

for $1 \leq k \leq n$, and

(5.200) $$y = (x_{n+1,n+1} + \tilde{x}_{n+1,n+1})/2 \otimes a,$$

where, again the matrices $x_{n+1,k}$ and $\tilde{x}_{n+1,k}$ were defined by (2.38) and (2.39) respectively.

We first state without proof of the obvious

Proposition 5.27: Let $(x,\tilde{x},y)$ be normalized system in C_1 which satisfies condition D (see Definition 5.20). Assume that there exists a matrix representation in which

(5.201) $$x = \begin{pmatrix} x_1 & 0 & 0 & 0 \\ 0 & x_2 & 0 & 0 \\ 0 & 0 & 0 & 0 \\ 0 & 0 & 0 & 0 \end{pmatrix} ; \quad \tilde{x} = \begin{pmatrix} 0 & 0 & 0 & 0 \\ 0 & 0 & 0 & 0 \\ 0 & 0 & x_1 & 0 \\ 0 & 0 & 0 & x_2 \end{pmatrix}$$

(5.202) $$y = \begin{pmatrix} 0 & 0 & y_1 & 0 \\ 0 & 0 & 0 & y_2 \\ y_3 & 0 & 0 & 0 \\ 0 & y_4 & 0 & 0 \end{pmatrix}$$

where the components x_j, $\tilde{x}_j$ and y_j do not vanish.

Then

(5.203) $$\|x_1\|_1 = \|y_1\|_1 = \|y_3\|_1 = \|\tilde{x}_1\|_1 = \lambda$$

(5.204) $$\|x_2\|_1 = \|y_2\|_1 = \|y_4\|_1 = \|\tilde{x}_2\|_1 = \mu$$

and each of the four systems satisfies condition M:

(5.205) $\quad (\lambda^{-1}x_1, \lambda^{-1}y_1;\ \lambda^{-1}\tilde{x}_1, \lambda^{-1}y_3)$

(5.206) $\quad (\mu^{-1}x_2, \mu^{-1}y_2; \mu^{-1}\tilde{x}_2, \mu^{-1}y_4)$

(5.207) $\quad (x, 2(y_1 + y_2);\ \tilde{x},\ 2(y_3 + y_4))$

(5.208) $\quad (x, 2(y_1 + y_4);\ \tilde{x},\ 2(y_2 + y_3))$

Lemma 5.28: Let $(z_1,\dots,z_n;\tilde{z}_1,\dots,\tilde{z}_n;y)$ be a normalized system in C_1 which satisfies condition DSM. Then there exist partial isometries $\{u_\alpha\}_{\alpha\in\mathcal{P}_{n+1,(0)}}$ and $\{w_\alpha\}_{\alpha\in\mathcal{P}_{n+1,(0)}}$ and an element $a \in C_1$ with $\|a\|_1 = 2^{1-n}$ so that

(5.209) $\quad r(u_\alpha) = \ell(a) = u_{\{1\}};\ r(w_\alpha) = r(a) = w_\phi\ ;$

(5.210) $\quad \ell(u_\alpha) = \ell_\alpha$ are pairwise orthogonal;

(5.211) $\quad \ell(w_\alpha) = r_\alpha$ are pairwise orthogonal;

and we have decompositions

(5.212) $$z_k = \sum_{\alpha\in\mathcal{P}^k_{n+1,(0)}} z_{k,\alpha}\ ;\quad \tilde{z}_k = \sum_{\alpha\in\mathcal{P}^k_{n+1,(e)}} \tilde{z}_{k,\alpha}$$

(5.213) $$y = \frac{1}{2}\sum_{\alpha\in\mathcal{P}^{n+1}_{n+1,(0)}} z_{n+1,\alpha} + \frac{1}{2}\sum_{\alpha\in\mathcal{P}^{n+1}_{n+1,(e)}} \tilde{z}_{n+1,\alpha}$$

where the components $z_{k,\alpha}$ and $\tilde{z}_{k,\alpha}$ $1 \leq k \leq n+1$, satisfy:

(5.214) $$z_{k,\alpha} = (-1)^{i(\alpha,k)} u_\alpha\ a\ w^*_{\alpha\sim\{k\}}$$

(5.215) $$\tilde{z}_{k,\alpha} = (-1)^{i(\alpha,k)} u_{\alpha\sim\{k\}}\ a\ w^*_\alpha$$

Proof: The system $(z_1,\dots,z_n;\tilde{z}_1,\dots,\tilde{z}_n)$ satisfies SM, and thus also the conclusion of Lemma 5.25. Using the fact that $\alpha \to \alpha \cup \{n+1\}$ is a one-to-one mapping of $\mathcal{P}_{n,(e)}$ onto $\mathcal{P}^{n+1}_{n+1,(0)}$ and of $\mathcal{P}_{n,(0)}$ onto $\mathcal{P}^{n+1}_{n+1,(e)}$, we can write the conclusion of Lemma 5.25 as follows.

There exist elements $a, \tilde{a} \in C_1$ with $a \perp \tilde{a}$ and $\|a + \tilde{a}\|_1 = 2^{2-n}$, there exist partial isometries

$$(5.216)\qquad \{u_\alpha\}_{\alpha\in \mathcal{P}_{n,(0)}},\ \{\tilde{u}_\alpha\}_{\alpha\in\mathcal{P}^{n+1}_{n+1,(0)}};\ \{w_\alpha\}_{\alpha\in \mathcal{P}_{n,(e)}},\ \{\tilde{w}_\alpha\}_{\alpha\in\mathcal{P}^{n+1}_{n+1,(e)}}$$

with

$$(5.217)\qquad r(u_\alpha) = \ell(a) = u_{\{1\}};\quad r(\tilde{u}_\alpha) = \ell(\tilde{a}) = \tilde{u}_{\{n+1\}};$$

$$(5.218)\qquad r(w_\alpha) = r(a) = w_\phi;\quad r(\tilde{w}_\alpha) = r(\tilde{a}) = \tilde{w}_{\{1,n+1\}};$$

$$(5.219)\qquad \{\ell(w_\alpha)\}_{\alpha\in \mathcal{P}_{n,(0)}} \cup \{\ell(\tilde{u}_\alpha)\}_{\alpha\in\mathcal{P}^{n+1}_{n+1,(0)}} \text{ are pairwise orthogonal;}$$

$$(5.220)\qquad \{\ell(w_\alpha)\}_{\alpha\in\mathcal{P}_{n,(e)}} \cup \{\ell(\tilde{u}_\alpha)\}_{\alpha\in\mathcal{P}^{n+1}_{n+1,(e)}} \text{ are pairwise orthogonal;}$$

and there exist decompositions

$$(5.221)\qquad z_k = \sum_{\alpha\in\mathcal{P}^k_{n+1,(0)}} z_{k,\alpha};\quad \tilde{z}_k = \sum_{\alpha\in\mathcal{P}^k_{n+1,(e)}} \tilde{z}_{k,\alpha}$$

for $1 \leq k \leq n$, where the components $z_{k,\alpha}$ and $\tilde{z}_{k,\alpha}$ satisfy

$$(5.222)\qquad z_{k,\alpha} = (-1)^{i(\alpha,k)} \cdot u_\alpha\, a\, w^*_{\alpha\sim\{k\}};\quad \alpha \in\mathcal{P}^k_{n,(0)}$$

$$(5.223)\qquad z_{k,\alpha} = (-1)^{i(\alpha,k)} \cdot \tilde{u}_\alpha\, \tilde{a}\, \tilde{w}^*_{\alpha\sim\{k\}};\quad \alpha \in\mathcal{P}^k_{n+1,(0)} \cap \mathcal{P}^{n+1}_{n+1}$$

$$(5.224)\qquad \tilde{z}_{k,\alpha} = (-1)^{i(\alpha,k)}\, u_{\alpha\sim\{k\}}\, a\, w^*_\alpha;\quad \alpha \in\mathcal{P}^k_{n,(e)}$$

(5.225) $\tilde{z}_{k,\alpha} = (-1)^{i(\alpha,k)}\tilde{u}_{\alpha\sim\{k\}}\,\tilde{a}\,\tilde{w}^*_\alpha;\quad \alpha \in \mathcal{P}^k_{n+1,(e)} \cap \mathcal{P}^{n+1}_{n+1}$.

Let us put also for $\alpha \in \mathcal{P}_{n+1,(0)}$ and $\beta \in \mathcal{P}_{n+1,(e)}$

(5.226)
$$\ell_\alpha = \begin{cases} \ell(u_\alpha) ; & n+1 \notin \alpha \\ \ell(\tilde{u}_\alpha) ; & n+1 \in \alpha \end{cases} \quad ; \quad r_\beta = \begin{cases} \ell(w_\beta) ; & n+1 \notin \beta \\ \ell(\tilde{w}_\beta) ; & n+1 \in \beta \end{cases}$$

Since $y = (\prod_{k=1}^{n} G_k \tilde{G}_k)y$, we have

(5.227) $y = (z_{n+1} + \tilde{z}_{n+1})/2$

where

(5.228)
$$z_{n+1} = \sum_{\alpha\in\mathcal{P}^{n+1}_{n+1,(0)}} z_{n+1,\alpha} \; ; \; \tilde{z}_{n+1} = \sum_{\alpha\in\mathcal{P}^{n+1}_{n+1,(e)}} \tilde{z}_{n+1,\alpha}$$

and for $\alpha \in \mathcal{P}^{n+1}_{n+1,(0)}$ and $\beta \in \mathcal{P}^{n+1}_{n+1,(e)}$ we have

(5.229) $z_{n+1,\alpha} = \ell_\alpha \cdot y = y \cdot r_{\alpha\sim\{n+1\}}$;

(5.230) $\tilde{z}_{n+1,\beta} = \ell_{\beta\sim\{n+1\}} \cdot y = y\cdot r_\beta$.

Now, by Proposition 5.27, we have $\|a\|_1 = \|\tilde{a}\|_1 = 2^{1-n}$ and the system $(z_1,\dots,z_n,z_{n+1};\tilde{z}_1,\dots,\tilde{z}_n,\tilde{z}_{n+1})$ satisfies SM. Thus, in particular

(5.231) $w = v(\tilde{z}_{n+1,\{1,n+1\}})^*\, v(\tilde{a}) = -\, v(\tilde{a})^* v(z_{n+1,\{n+1\}})$

(5.232) $u = v(\tilde{a})v(\tilde{z}_{n+1,\{1,n+1\}})^* = -\, v(z_{n+1,\{n+1\}})v(a)^*$

(5.233) $z_{n+1,\{n+1\}} = -\, ua,\ \tilde{z}_{n+1,\{1,n+1\}} = aw^*$

(5.234) $\tilde{a} = u\, a\, w^*$

Let us define for $\alpha \in \mathcal{P}^{n+1}_{n+1,(0)}$ and $\beta \in \mathcal{P}^{n+1}_{n+1,(e)}$

(5.235) $\quad u_\alpha = \tilde{u}_\alpha u \ ; \quad w_\beta = \tilde{w}_\beta w$

Then, clearly, the systems $\{u_\alpha\}_{\alpha \in \mathcal{P}_{n+1,(0)}}$ and $\{w_\beta\}_{\beta \in \mathcal{P}_{n+1,(e)}}$ satisfy (5.209), (5.210) and (5.211), we have decompositions (5.212) and (5.213), and (5.214) and (5.215) are satisfied for $1 \leq k \leq n$ and every α, and for $k = n+1$ and $\alpha = \{n+1\}$ and α $\{1,n+1\}$ respectively. From this fact we get as in the proof of Lemma 5.25 that (5.214) and (5.215) hold for $k = n+1$ and every α, using the fact that $(z_1,\ldots,z_{n+1};\tilde{z}_1,\ldots,\tilde{z}_{n+1})$ satisfies SM. □

Corollary 5.29: (i) Let $(z_1,\ldots,z_n;\tilde{z}_1,\ldots,\tilde{z}_n;y)$ be a normalized system in C_1 which satisfies condition DSM. Then there exists an element $a \in C_1$ with $\|a\|_1 = 2^{1-n}$, and there exists a tensor product representation in which:

(5.236) $\quad z_k = x_{n+1,k} \otimes a; \quad \tilde{z}_k = \tilde{x}_{n+1,k} \otimes a; \quad 1 \leq k \leq n$

(5.237) $\quad y = (x_{n+1,n+1} + \tilde{x}_{n+1,n+1})/2 \otimes a$

where the matrices $x_{n+1,k}$ and $\tilde{x}_{n+1,k}$ were defined by (2.38) and (2.39) respectively. In particular, there exists a contractive projection from C_1 onto $\text{span}\{z_1,\ldots,z_n;\tilde{z}_1,\ldots,\tilde{z}_n;y\} = DAH_1^{n+1}(a)$, which is canonical in the sense of §2d.

(ii) Let P be a contractive projection in C_1 so that $X = R(P)$ is irreducible. Assume that case 3 of Proposition 5.3 occurs and the case "$\dim R(Q_n) = 1$" of Lemma 5.23 occurs.

Then $X = \text{span}\{z_1,\ldots,z_n;\tilde{z}_1,\ldots,\tilde{z}_n;y\}$, where $(z_1,\ldots,z_n;\tilde{z}_1,\ldots,\tilde{z}_n;y)$ is a normalized system which satisfies DSM, $PG(X) = 0$ and $PE(X)$ is the canonical contractive projection from C_1 onto X. So, X is an elementary subspace of C_1 of type 6, i.e. a space of the form $DAH_1^{n+1}(a)$.

The corollary follows by Lemmas 5.23 and 5.28; the proof is very similar to the proof of Corollary 5.26, and thus we omit it.

(g) - Proof of theorem 2.14, 2.15 and 2.16

Proof of Theorem 2.14: Let P be a contractive projection in C_1, and let $X = R(P)$. By Proposition 5.2 there exists a family $\{X_k\}_{k\in K}$, $1 \leq |K| \leq \aleph_0$, of subspaces of X which are irreducible and are pairwise disjointly supported, and so that $X = \sum_{k\in K} X_k$. Also, $P_k = E(X_k)P$ is a contractive projection from C_1 onto X_k and $Px = \sum_{k\in K} P_k x$ for every $x \in C_1$ (where, if $|K| = \aleph_0$ the series $\sum_{k\in K} P_k x$ converges absolutely). Thus, Theorem 2.14 reduces to the following proposition.

Proposition 5.30: Let P be a contractive projection in C_1 and assume that $X = R(P)$ is irreducible. Then X is an elementary subspace of C_1.

Proof: Let x be an atom of P. By Proposition 5.3 one (and only one) of the mutually exclusive cases 1-5 occurs.

In case 1 we get by Lemma 5.4 and Remark 5.6 that $X = SY_1^m(x)$, so X is an elementary subspace of type 2.

In case 2 we get by the work in subsection 5e that $X = H_1^n(y_1,\dots,y_n)$, so X is an elementary subspace of type 2.

In case 3 we get by the work in subsection 5f that either $X = AH_1^n(a,\tilde{a})$ or $X = DAH_1^n(a)$ for some $n < \infty$ and elements $a,\tilde{a} \in C_1$. So, X is an elementary of type 5 or type 6 respectively.

In case 4, we get by Lemma 5.9 and Remark 5.10 that $X = A_1^m(b)$ for some $5 \leq m \leq \infty$ and $b \in C_1$, so X is an elementary subspace of C_1 of type 3. Finally, in case 5 we get by Lemma 5.7 and Remark 5.8 that $X = C_1^{n,m}(a,b)$ for some $2 \leq n \leq m \leq \infty$, $3 \leq m$ and some $a,b \in C_1$. So, X is an elementary subspace of C_1 of type 1.

This completes the proof of Proposition 5.30, and thus - completes the proof of Theorem 2.14. □

Proof of Theorem 2.15: Let P be a contractive projection in C_1 and let $X = R(P)$. Let $X = \sum_{k \in K} X_k$ be the decomposition of X into pairwise disjointly supported irreducible subspaces, so that each X_k is an elementary subspace of C_1, as described in Theorem 2.14. Put E_k, F_k and G_k for $E(X_k)$, $F(X_k)$ and $G(X_k)$ respectively. Then, the family $\{E_k, F_k, G_k\}_{k \in K}$ is commutative. By the work in this section (more precisely, by Lemmas 5.4, 5.7 and 5.23) we get that the contractive projection $P_k = E_k P$ from C_1 onto X_k satisfies $P_k G_k = 0$, and $Q_k = P_k E_k$ is the canonical contractive projection from C_1 onto the elementary subspace X_k, in the sense of §2.

Let $k, j \in K$ with $k \neq j$, and let $0 \neq x_j \in X_j$. By Lemma 3.1 we have

$$PE(x_j) = E(x_j)P\, E(x_j),\ F(x_j)P\, G(x_j) = 0 \tag{5.238}$$

So

$$P_k E(x_j) = E_k PE(x_j) = E_k E(x_j) PE(x_j) = 0 \tag{5.239}$$

$$P_k G(x_j) = E_k PG(x_j) = E_k F(x_j) PG(x_j) = 0 \tag{5.240}$$

Thus

$$P_k = P_k\, F(x_j) \tag{5.241}$$

Since $x_j \in X_j$ is arbitrary we deduce from (5.241) that

$$P_k = P_k F_j \tag{5.242}$$

and so

$$\begin{aligned} P_k &= P_k \Big(\prod_{\substack{j \in K \\ j \neq k}} F_j \Big) = P_k (E_k + G_k + F_k) \Big(\prod_{\substack{j \in K \\ j \neq k}} F_j \Big) \\ &= P_k E_k + P_k \Big(\prod_{j \in K} F_j \Big) = Q_k + P_k\, F(X) \end{aligned} \tag{5.243}$$

Let $Q = \sum_{k\in K} Q_k$ and $T = P - Q$. Then, clearly, Q is a contractive projection from C_1 onto X which satisfies $Q = QE(X) = PE(X)$, and

$$(5.244)\qquad T = P - Q = \sum_{k\in K} P_k - \sum_{k\in K} Q_k = \sum_{k\in K} P_k\, F(X) = PF(X).$$

It follows that

$$(5.245)\qquad T^2 = 0,\ \|T\| \leq 1,\ TF(X) = T = QT = PT.$$

Conversely, let $\{X_k\}_{k\in K}$ be a family of pairwise disjointly supported elementary subspaces of C_1 and let Q_k be the canonical contractive projection from C_1 onto X_k in the sense of §2. Put $Q = \sum_{k\in K} Q_k$, let T be an operator on C_1 which satisfies

$$(5.246)\qquad T^2 = 0,\ \|T\| \leq 1,\quad TF(X) = T = QT,$$

and let $P = Q + T$. Then

$$(5.247)\qquad P^2 = Q^2 + QT + TQ + T^2 = Q + T + TF(X)Q + 0 = P$$

$$(5.248)\qquad P = Q + T = Q(1 + T)$$

$$(5.249)\qquad PT = (Q + T)T = QT + T^2 = T$$

$$(5.250)\qquad Q = P - T = P(1 - T)$$

So P is a projection with $R(P) = R(Q) = X$. Also, for every $x \in C_1$

$$(5.251)\qquad \|Px\|_1 = \|(Q + T)x\|_1 \leq \|Qx\|_1 + \|Tx\|_1$$

$$= \|QE(X)x\|_1 + \|TF(X)x\|_1$$

$$\leq \|E(X)x\|_1 + \|F(X)x\|_1 \leq \|x\|_1$$

Thus, P is also contractive. □

Remark 5.31: Let $X, \{X_k\}_{k\in K}, Q_k$ and $Q = \sum_{k\in K} Q_k$ as in the proof of Theorem 2.15. Then $Q = QE(X)$, and if P is any contractive projection from C_1 onto X then $PE(X) = Q$. Thus Q is the unique contractive projection from C_1 onto X which satisfies $QE(X) = Q$. Thus Q is called the "canonical" (or "minimal") contractive projection from C_1 onto X. Also, since the elementary subspaces $\{X_k\}_{k\in K}$ are disjointly supported, X is isometric to $(\sum_{k\in K} \oplus X_k)_{\ell_1}$.

Theorems 2.14 and 2.15 describe completely the contractive projections and their ranges in C_1. By duality, we get the complete description of contractive projections and their ranges in C_∞

Proof of Theorem 2.16: Let P be a contractive projection in C_∞, and let $X = R(P)$. Since P^* is a contractive projection in C_1 we get by Theorems 2.14 and 2.15 that $R(P^*) = Z = \sum_{k\in K} Z_k$, $1 \leq |K| \leq \aleph_0$, where $\{z_k\}_{k\in K}$ are pairwise disjointly supported subspaces of Z which are elementary subspaces of C_1, and if $\hat{Q}_k$ is the canonical projection from C_1 onto Z_k then $P^* = \hat{Q} + \hat{T}$, where $\hat{Q} = \sum_{k\in K} \hat{Q}_k$ and

$$(2.252) \qquad \|\hat{T}\| \leq 1,\ \hat{T}^2 = 0,\ \hat{T} = \hat{Q}\hat{T} = \hat{T}F(Z)$$

Also, $\hat{Q} = \hat{Q}E(Z)$ and thus $\hat{Q}^* = E(Z)\hat{Q}^*$. Since by (2.252) we have $\hat{T}^* = F(Z)\hat{T}^*$ we get that for any $y \in B(\ell_2), \hat{T}^* y \perp \hat{Q}^* y$. In particular, if $y \in C_\infty$ then

$$(5.253) \qquad P^{**}y = Py \in C_\infty,\ Py = \hat{Q}^*y + \hat{T}^*y,\ \hat{Q}^*y \perp \hat{T}^*y.$$

This implies that $\hat{Q}^*y$ and $\hat{T}^*y$ belongs to C_∞, and thus $\hat{Q}$ and $\hat{T}$ are adjoint operators (being w^*-continuous). Let Q and T be operators on C_∞ so that $Q^* = \hat{Q}$ and $T^* = \hat{T}$. An easy check yields that $Q = \sum_{k\in K} Q_k$, where Q_k is a canonical contractive projection from C_∞ onto an elementary subspace Y_k, $Q_k^* = \hat{Q}_k$, and $\{Y_k\}_{k\in K}$ are pairwise disjointly supported. If

$\{z_j^{(k)}\}_{j\in A_k} \subset R(\hat{Q}_k)$ is an appropriate normalized system so that $Z_k = \overline{\text{span}}\ \{z_j^{(k)}\}_{j\in A_k}$ and for every $x \in C_1$

$$(5.254) \quad \hat{Q}_k x = \sum_{j\in A_k} \langle x, v(z_j^{(k)})\rangle\, z_j^{(k)}$$

(where if A_k is infinite, the series (5.254) converges in the norm of C_1 in some ordering), then the $z_j^{(k)}$ are of finite rank, and for every $y \in C_\infty$

$$(5.255) \quad Q_k y = \sum_{j\in A_k} \langle y, z_j^{(k)}\rangle\, v(z_j^{(k)})$$

(where, again, if A_k is infinite, the series (5.255) converges in the norm of C_∞ in the above ordering).

Also, by (5.252) we get that

$$(5.256) \quad \|T\| \leq 1, \quad T^2 = 0, \ T = TQ = F(Z)T$$

and since $P^* = \hat{Q} + \hat{T} = Q^* + T^*$ we get

$$(5.257) \quad P = Q + T.$$

Put

$$(5.258) \quad Y = \sum_{k\in K} Y_k = R(Q)$$

and for every $k\in K$ let

$$(5.259) \quad X_k = (1 + T)Y_k \ .$$

If $x \in C_\infty$, then

$$(5.260) \quad Px = (1 + T)Qx = \sum_{k\in K} (1 + T)Q_k\, x$$

and if $x \in X$ and $Q_k x = y_k \in Y_k$ and $x_k = (1 + T)y_k \in X_k$, then

$$(5.261)\quad \|x\|_\infty = \|\sum_{k\in K} y_k + T\sum_{k\in K} y_k\|_\infty$$

$$= \max\{\|\sum_{k\in K} y_k\|_\infty, \|T\sum_{k\in K} y_k\|_\infty\}$$

$$= \|\sum_{k\in K} y_k\|_\infty = \max_{k\in K} \|y_k\|$$

$$= \max_{k\in K} \|x_k\| .$$

Thus

$$(5.262)\quad X = \sum_{k\in K} X_k$$

and the operator $1 + T$ is an isometry from Y onto X. So X is isometric to

$$(5.263)\quad (\sum_{k\in K} \oplus Y_k)_{c_0} .$$

Conversely, let $\{Y_k\}_{k\in K}$ $1 \leq |K| \leq \aleph_0$ be a family of pairwise disjointly supported elementary subspaces of C_∞, let Q_k be the canonical contractive projection from C_∞ onto Y_k and let $Q = \sum_{k\in K} Q_k$ be the canonical contractive projection from C_∞ onto $Y = \sum_{k\in K} Y_k$. Let T be an operator on C_∞ so that $\|T\| \leq 1$, $T^2 = 0$ and $T = TQ = F(Y)T$. Put $P = Q + T$. Since $Q = Q\,E(Y)$, $QT = 0$. Thus $P^2 = P$. Also, for every $y \in C_\infty$, $Qy \perp Ty$, so

$$(5.264)\quad \|Py\|_\infty = \|Qy + Ty\|_\infty = \max\{\|Qy\|_\infty, \|Ty\|_\infty\} \leq \|y\|_\infty$$

Thus P is a contractive projection, and as before $X = R(P)$ is isometric to Y. $\square$

§6. ISOMETRIES FROM ELEMENTARY SUBSPACES OF C_1 INTO C_1.

The main results of this section are the following theorems:

Theorem 6.1 Let X be an elementary subspace of C_1 of type different from 4, and let V be an isometry from X into C_1. Then $V(X)$ is an elementary subspace of C_1 of the same type as X. In particular, there exists a contractive projection from C_1 onto $V(X)$.

Theorem 6.2 Let V be an isometry from a Hilbert space H into C_1 so that

$$\{x,y \in H,\quad x \neq 0 \neq y,\quad (x,y) = 0\} \Rightarrow (Vx)G(Vy). \tag{6.1}$$

Then $V(H)$ is an elementary subspace of C_1 of type 4, i.e. an elementary Hilbert subspace. In particular, there exists a contractive projection from C_1 onto $V(H)$.

Recall that a normalized system $(x_1,x_2;\tilde{x}_1,\tilde{x}_2)$ of elements of C_1 satisfies the matrix condition (condition "M", see also Definition 4.4 and Proposition 4.5), if there exists a tensor product representation and elements $a,b \in C_1$ with $a \perp b$ and $\|a+b\|_1 = 1$ so that

$$(6.2)\quad \begin{cases} x_1 = e_{1,1} \otimes a + e_{1,1} \otimes b; & x_2 = e_{1,2} \otimes a + e_{2,1} \otimes b \\ \tilde{x}_1 = e_{2,2} \otimes a + e_{2,2} \otimes b; & \tilde{x}_2 = -(e_{2,1} \otimes a + e_{1,2} \otimes b) \end{cases}$$

Also, a normalized system $(x,\tilde{x},y)$ of elements of C_1 satisfies the diagonal condition (condition "D", see also Definition 5.20) if there exists a tensor product representation and an element $a \in C_1$ with $\|a\|_1 = 1$ so that

$$x = e_{1,1} \otimes a,\quad \tilde{x} = -e_{2,2} \otimes a,\quad y = 1/2(e_{1,2}+e_{2,1}) \otimes a \tag{6.3}$$

For the proofs of Theorems 6.1 and 6.2 we need the following tools:

Lemma 6.3 [1, Theorem 2.1] Let $x_1,x_2,\tilde{x}_1,\tilde{x}_2$ be normalized elements in C_1. Then the system $(x_1,x_2;\tilde{x}_1,\tilde{x}_2)$ satisfies the matrix condition if and only if for every choice of scalars $t_1,t_2,\tilde{t}_1,\tilde{t}_2$ we have:

$$\|t_1x_1 + t_2x_2 + \tilde{t}_1\tilde{x}_1 + \tilde{t}_2\tilde{x}_2\|_1 = \left\|\begin{pmatrix} t_1 & t_2 \\ -\tilde{t}_2 & \tilde{t}_1 \end{pmatrix}\right\|_1 \tag{6.4}$$

Thus, if $(x_1,x_2;\tilde{x}_1,\tilde{x}_2)$ satisfies the matrix condition and if V is an isometry from $\mathrm{span}\{x_1,x_2,\tilde{x}_1,\tilde{x}_2\}$ into C_1, then also $(Vx_1, Vx_2; V\tilde{x}_1, V\tilde{x}_2)$ satisfies the matrix condition.

<u>Lemma 6.4</u> Let $x,\tilde{x},y$ be normalized elements in C_1. Then the system $(x,\tilde{x},y)$ satisfies the diagonal condition if and only if for every choice of scalars $t,\tilde{t},s$ we have:

$$\|tx + \tilde{t}\tilde{x} + sy\|_1 = \left\|\begin{pmatrix} t & s/2 \\ s/2 & -\tilde{t} \end{pmatrix}\right\|_1 \tag{6.5}$$

Thus, if $(x,\tilde{x},y)$ satisfies the diagonal condition, and if V is an isometry from $\mathrm{span}\{x,\tilde{x},y\}$ into C_1, then also $(Vx,V\tilde{x},Vy)$ satisfies the diagonal condition.

In both lemmas the "only if" part, namely that the matrix and the diagonal condition imply formulas (6.4) and (6.5) respectively, is trivial and follows directly from the definitions. The "if" part of Lemma 6.3 is proved in [1] in the more general context of isometries of $C_p^{n,m}$ $(1 \leq p \neq 2 < \infty,\ 2 \leq n,m \leq \infty)$ into C_p.

<u>Proof of the "if" part in Lemma 6.4:</u> Let $x,\tilde{x},y$ be elements of C_1 such that (6.5) holds for any choice of scalars t, $\tilde{t}$ and s. Let us define an operator $V\colon SY_1^2 \to \mathrm{span}\{x,\tilde{x},y\}$ by $Ve_{1,1} = x$, $Ve_{2,2} = -\tilde{x}$ and $Vs_{1,2} = 2y$. By (6.5), V is an isometry, and thus by Proposition 1.1 V preserves the relation "$\perp$" of the disjointness of the supports. If $\lambda > 0$, then it is easily verified that

$$(6.6)\qquad \begin{pmatrix} \lambda^2 & \lambda \\ \lambda & 1 \end{pmatrix} \perp \begin{pmatrix} 1 & -\lambda \\ -\lambda & \lambda^2 \end{pmatrix}$$

and so

$$(6.7)\qquad (\lambda^2 x + 2\lambda y - \tilde{x}) \perp (x - 2\lambda y - \lambda^2\tilde{x})$$

Similarly, since $e_{1,1} \perp e_{2,2}$, we get that

$$(6.8)\qquad x \perp \tilde{x}$$

From (6.7) and (6.8) it follows that for every $\xi \in \ell_2$

$$\begin{aligned}(6.9)\qquad 0 &= ((\lambda^2 x + 2\lambda y - \tilde{x})\xi, (x - 2\lambda y - \lambda^2\tilde{x})\xi) \\ &= 2\lambda[(y\xi, x\xi) + (\tilde{x}\xi, y\xi)] \\ &+ \lambda^2[\|x\xi\|^2 + \|\tilde{x}\xi\|^2 - 4\|y\xi\|^2] \\ &- 2\lambda^3[(x\xi, y\xi) + (y\xi, \tilde{x}\xi)]\end{aligned}$$

(Here, $(\cdot,\cdot)$ is the inner product in ℓ_2, and $\|\cdot\|$ is the corresponding inner product norm).

If $x\xi = \tilde{x}\xi = 0$, then by (6.9) we have also $y\xi = 0$. Also, (6.7) and (6.8) imply that (6.9) is true also with $x^*, \tilde{x}^*$, and y^* instead of x, $\tilde{x}$ and y. So, $x^*\xi = \tilde{x}^*\xi = 0$ implies that $y^*\xi = 0$. Hence $y = E(x+\tilde{x})y$, and we have a matrix representation in which

$$(6.10)\qquad x = \begin{pmatrix} x & 0 \\ 0 & 0 \end{pmatrix}; \quad \tilde{x} = \begin{pmatrix} 0 & 0 \\ 0 & \tilde{x} \end{pmatrix}; \quad y = \begin{pmatrix} y_{1,1} & y_{1,2} \\ y_{2,1} & y_{2,2} \end{pmatrix}$$

Now if $0 \neq \xi \in \ell_2$ is such that $\tilde{x}\xi = 0$ but $r(x)\xi = \xi$, then for every $\lambda > 0$ we get by (6.9)

(6.11) $$0 = 2\lambda(y\xi,x\xi) + \lambda^2(\|x\xi\|^2 - 4\|y\xi\|^2) - 2\lambda^3(x\xi,y\xi)$$

So $(x\xi,y\xi) = (\xi,x^*y\xi) = 0$. Since the quadratic form $(\xi,x^*y\xi) = \psi(\xi)$ vanishes identically on $r(x)\ell_2$, its associate bilinear form must also vanish. So,

(6.12) $$y_{1,1} = E(x)y = 0.$$

Similarly,

(6.13) $$y_{2,2} = E(\tilde{x})y = 0$$

Using again (6.11) for any $0 \neq \xi \in \ell_2$ with $r(x)\xi = \xi$ and $\tilde{x}\xi = 0$, we get:

(6.14) $$(x^*x\xi,\xi) = \|x\xi\|^2 = 4\|y\xi\|^2 = 4(y^*y\xi,\xi) = 4(y_{2,1}^*y_{2,1}\xi,\xi)$$

By the polarization identity in ℓ_2 we get from (6.14) that

(6.15) $$|x|^2 = |2y_{2,1}|^2, \quad |x| = |2y_{2,1}|$$

Using the pairs $\{\tilde{x},y\}$, $\{x^*,y^*\}$ and $\{\tilde{x}^*,y^*\}$ we get in the same way that also

(6.16) $$|\tilde{x}| = |2y_{1,2}|, \quad |x^*| = |2y_{1,2}^*|, \quad |\tilde{x}^*| = |2y_{2,1}^*|$$

Put

(6.17) $$w = v(y)^*v(x) = v(y_{1,2})^*v(x)$$

(6.18) $$u = v(y)v(x)^* = v(y_{2,1})v(x)^*$$

Then, by (6.15) and (6.16)

(6.19) $$y = (ux + xw^*)/2$$

Also, by Proposition 1.2, (6.15) and (6.16)

$$(6.20)\qquad 4 = \left\|\begin{pmatrix}1 & 1\\ 1 & 1\end{pmatrix}\right\|_1^2 = \|x + 2y - \tilde{x}\|_1^2 = \left\|\begin{pmatrix}x & 2y_{1,2}\\ 2y_{2,1} & -\tilde{x}\end{pmatrix}\right\|_1^2 .$$

$$\geq \left\|\begin{pmatrix}x & 0\\ 2y_{2,1} & 0\end{pmatrix}\right\|_1^2 + \left\|\begin{pmatrix}0 & 2y_{1,2}\\ 0 & -\tilde{x}\end{pmatrix}\right\|_1^2 = 2 + 2 = 4.$$

So we have an equality everywhere in (6.20). By the equality case in Proposition 1.2 there exists a partial isometry σ with $r(\sigma) = r(x)$ and $\ell(\sigma) = r(\tilde{x})$ so that

$$(6.21)\qquad (x + 2y_{2,1})\sigma^* = 2y_{1,2} - \tilde{x}$$

Multiplying (6.21) on the left by $\ell(x)$, we get

$$(6.22)\qquad x\sigma^* = 2y_{1,2}$$

So (6.16) and (6.19) imply that $\sigma = w$. Multiplying (6.21) on the left by $\ell(\tilde{x})$ we get

$$(6.23)\qquad 2y_{2,1}\sigma^* = -\tilde{x}$$

Thus

$$(6.24)\qquad \tilde{x} = -2y_{2,1}w^* = -uxw^*$$

Clearly, (6.19) and (6.24) imply that the system $(x,\tilde{x},y)$ satisfies the diagonal condition.

<u>Proof of Theorem 6.1</u> Let X be an elementary subspace of C_1 of type different from 4, and let V be an isometry from X into C_1. The

proof that $V(X)$ is also an elemetnary subspace of C_1 (of the same type of X) depends on the type of X.

If X is of type 1, i.e. $X = C_1^{n,m}(a,b)$ $(2 \leq n \leq m \leq \infty,\ 3 \leq m)$ in some tensor product prepresentation, then by Proposition 2.2 and by [1, Theorem 2.1] we get that in some (generally, different) tensor product representation we have $V(X) = C_1^{n,m}(\hat{a},\hat{b})$, and thus $V(X)$ is an elementary subspace of type 1.

If X is of type 5, then $X = \text{span}\{x_k, \tilde{x}_k\}_{k=1}^n$ $2 \leq n < \infty$, where $(x_1,\dots,x_n;\ \tilde{x}_1,\dots,\tilde{x}_n)$ is a normalized system in C_1 which satisfies SM (see §2c, and Definition 5.21). Let $z_k = Vx_k$ and $\tilde{z}_k = V\tilde{z}_k$. By Lemma 6.3 V preserves the matrix condition, and thus V preserves also condition SM. It follows that the system $(z_1,\dots,z_n;\ \tilde{z}_1,\dots,\tilde{z}_n)$ satisfies SM. By Corollary 5.26 there exist elements a, $\tilde{a} \in C_1$ and there exists a tensor-product representation so that (5.192) and (5.193) hold. Thus

$$(6.25) \qquad V(X) = \text{span}\{z_k, \tilde{z}_k\}_{k=1}^n = AH_1^n(a,\tilde{a})$$

and so $V(X)$ is an elementary subspace of C_1 of type 5.

If X is of type 6, then $X = \text{span}\{x_1,\dots,x_n;\tilde{x}_1,\dots,\tilde{x}_n;u\}$, $2 \leq n < \infty$, where $(x_1,\dots,x_n;\tilde{x}_1,\dots,\tilde{x}_n;u)$ is a normalized system in C_1 which satisfies DSM(see §2d, and Definitions 5.20 and 5.21). Put $z_k = Vx_k$, $\tilde{z}_k = V\tilde{x}_k$ and $y = Vu$. Since by Lemmas 6.3 and 6.4 V preserves conditions M and D, we get from Definitions 5.21 that V also preserves condition DSM. Thus $(z_1,\dots,z_n;\tilde{z}_1,\dots,\tilde{z}_n;y)$ satisfies DSM. By Corollary 5.29 there exist a tensor-product representation and an element $a \in C_1$ so that formulas (5.236) and (5.237) hold. So

$$(6.26) \qquad V(X) = \text{span}\{z_1,\dots,z_n;\tilde{z}_1,\dots,\tilde{z}_n;y\} = DAH_1^{n+1}(a)$$

and thus $V(X)$ is an elementary subspace of C_1 of type 6.

<u>If X is of type 2,</u> then $X = \overline{\mathrm{span}}\{x_{i,j}\}_{1\le i\le j\le n}$, $n \le \infty$ where for every choice of scalars $\{t_{i,j}\}_{1\le i\le j<n}$ we have

$$(6.27)\qquad \Big\| \sum_{1\le i\le j\le n} t_{i,j}x_{i,j} \Big\|_1 = \Big\| \sum_{1\le i\le j\le n} t_{i,j}s_{i,j} \Big\|_1$$

where $s_{i,j}$ are the elementary symmetric matrices, i.e. $s_{i,i} = e_{i,i}$ and $s_{i,j} = e_{i,j} + e_{j,i}$ for $i < j$. Put $y_{i,j} = Vx_{i,j}$, $1 \le i \le j \le n$. Then for every choice of scalars $\{t_{i,j}\}_{1\le i\le j\le n}$ we have:

$$(6.28)\qquad \Big\| \sum_{1\le i\le j\le n} t_{i,j}y_{i,j} \Big\|_1 = \Big\| \sum_{1\le i\le j\le n} t_{i,j}s_{i,j} \Big\|_1$$

By Lemma 6.4 we get that for every $1 \le i < j \le n$

$$(6.29)\qquad (y_{i,i}, -y_{j,j}, y_{i,j}/2) \text{ satisfies } D.$$

It follows that $y_{i,i} \perp y_{j,j}$ for $i \ne j$, and that for $1 \le i < j \le n$ $y_{i,j} = b_{i,j} + c_{i,j}$ where

$$(6.30)\qquad b_{i,j} = \ell(y_{i,i})y_{i,j}r(y_{j,j}); \quad c_{i,j} = \ell(y_{j,j})y_{i,j}r(y_{i,i})$$

and

$$(6.31)\qquad |b_{i,j}| = |y_{j,j}|, \quad |b_{i,j}^*| = |y_{i,i}^*|;$$

$$(6.32)\qquad |c_{i,j}| = |y_{i,i}|, \quad |c_{i,j}^*| = |y_{j,j}^*|.$$

Put

$$(6.33)\qquad y = y_{1,1}, \quad u_1 = \ell(y), \quad w_1 = r(y)$$

and for $1 < i \le n$,

$$(6.34)\qquad w_i = v(b_{1,i})^* v(y); \quad u_i = v(c_{1,i})v(y)^*$$

We claim that for every $1 \leq i < j \leq n$

$$(6.35) \qquad b_{i,j} = u_i y w_j^*, \qquad c_{i,j} = u_j y w_i^*$$

and for every $1 \leq i \leq n$,

$$(6.36) \qquad y_{i,i} = u_i y w_i^*$$

Since $(y_{1,1}, -y_{j,j}, \frac{1}{2}y_{1,j})$ satisfies D, $1 < j \leq n$, we get (6.35) for $i = 1$ and j, and (6.36) for $i = j$. Let now $1 < i < j \leq n$. By (6.28), (6.31) and (6.32) and Proposition 1.2 we get

$$(6.37) \qquad 9 = \left\| \begin{pmatrix} 1 & 1 & 1 \\ 1 & 1 & 1 \\ 1 & 1 & 1 \end{pmatrix} \right\|_1^2 = \left\| \begin{pmatrix} y & b_{1,i} & b_{1,j} \\ c_{1,i} & y_{i,i} & b_{i,j} \\ c_{1,j} & c_{i,j} & y_{j,j} \end{pmatrix} \right\|_1^2$$

$$\geq \left\| \begin{pmatrix} y & 0 & 0 \\ c_{1,i} & 0 & 0 \\ c_{1,j} & 0 & 0 \end{pmatrix} \right\|_1^2 + \left\| \begin{pmatrix} 0 & b_{1,i} & 0 \\ 0 & y_{1,i} & 0 \\ 0 & c_{i,j} & 0 \end{pmatrix} \right\|_1^2 + \left\| \begin{pmatrix} 0 & 0 & b_{1,j} \\ 0 & 0 & b_{i,j} \\ 0 & 0 & y_{i,j} \end{pmatrix} \right\|_1^2$$

$$= 3 + 3 + 3 = 9.$$

So there exists an equality in (6.37). By the equality case of Proposition 1.2 we get a partial isometry w with $r(w) = r(b_{1,i})$, $\ell(w) = r(b_{1,j})$ and so that

$$(6.38) \qquad (b_{1,j} + b_{i,j} + y_{j,j}) = (b_{1,i} + y_{i,i} + c_{i,j})w^*$$

Multiplying (6.38) on the left by $\ell(y)$ we get

(6.39) $b_{1,j} = b_{1,i}w^* = b_{1,i}w_i w_j^*$

Using (6.31) we deduce that

(6.40) $w = w_j w_i^*$

and thus, by multiplying (6.38) on the left by $\ell(y_{i,i})$,

(6.41) $b_{i,j} = y_{i,i}w^* = (u_i y w_i^*)w_i w_j^* = u_i y w_j^*$

Similarly, we get from the equality in (6.37) and the equality case in Proposition 1.2 that

(6.42) $c_{i,j} = u_i y w_i^*$

This completes the proof of (6.35).

From (6.35) and (6.36) we deduce the existence of a tensor product representation in which

(6.43) $y_{i,j} = s_{i,j} \otimes y$, $1 \leq i \leq j \leq n$

So $V(X) = \overline{\text{span}}\{y_{i,j}\}_{1\leq i\leq j\leq n} = SY_1^n(y)$, and thus $V(X)$ is an elementary subspace of C_1 of type 2.

<u>If X is of type 3</u>, then $X = \overline{\text{span}}\{x_{i,j}\}_{1\leq i<j\leq n}$, $5 \leq n \leq \infty$ and for every choice of scalars $\{t_{i,j}\}_{1\leq i<j\leq n}$ we have

(6.44) $\|\sum_{1\leq i<j\leq n} t_{i,j}x_{i,j}\|_1 = \|\sum_{1\leq i<j\leq n} t_{i,j}a_{i,j}\|_1$

where the $a_{i,j}$ are the elementary anti-symmetric matrices, i.e. $a_{i,j} = e_{i,j} - e_{j,i}$ for $1 \leq i < j \leq n$.

Put $y_{i,j} = Vx_{i,j}$, then for every choice of scalars $\{t_{i,j}\}_{1 \leq i<j \leq n}$ we have

$$(6.45) \qquad \Big\| \sum_{1\leq i<j\leq n} t_{i,j} y_{i,j} \Big\|_1 = \Big\| \sum_{1\leq i<j\leq n} t_{i,j} a_{i,j} \Big\|_1$$

In particular,

$$(6.46) \qquad \Big\| \sum_{\substack{1\leq i\leq 2\\ 3\leq j\leq n}} t_{i,j} y_{i,j} \Big\|_1 = \Big\| \sum_{\substack{1\leq i\leq 2\\ 3\leq j\leq n}} t_{i,j} a_{i,j} \Big\|_1 = 2\Big\| \sum_{\substack{1\leq i\leq 2\\ 3\leq j\leq n}} t_{i,j} e_{i,j} \Big\|_1$$

Since $5 \leq n$, we can use [1, Theorem 2.1] to get elements $b,c \in C_1$ with $b \perp c$ and $\|b+c\|_1 = 1$, and partial isometries u_i, w_i, $\tilde{u}_j$ and $\tilde{w}_j$, $i = 1,2$ $3 \leq j \leq n$, so that:

$$(6.47) \qquad r(u_2) = \ell(b) = u_1; \quad r(w_2) = r(c) = w_1$$

$$(6.48) \qquad r(\tilde{u}_j) = \ell(c) = \tilde{u}_3; \quad r(\tilde{w}_j) = r(b) = \tilde{w}_3$$

(6.49) the projections $\{\ell(u_i)\}_{i=1}^2 \cup \{\ell(\tilde{u}_j)\}_{j=3}^n$ are pairwise orthogonal;

(6.50) the projections $\{\ell(w_i)\}_{i=1}^2 \cup \{\ell(\tilde{w}_j)\}_{j=3}^n$ are pairwise orthogonal; and for $1 \leq i \leq 2$ and $3 \leq j \leq n$ we have

$$(6.51) \qquad y_{i,j} = b_{i,j} + c_{i,j}$$

$$(6.52) \qquad b_{i,j} = u_i b \tilde{w}_j^* ; \quad c_{i,j} = \tilde{u}_j \, c \, w_i^*$$

Now, if $1 \leq i_1 < i_2 < i_3 < i_4 \leq n$ and if we put

(6.53) $$z_{k,j} = y_{i_k,i_j}/2, \quad 1 \leq k < j \leq 4$$

then by (6.45) and Lemma 6.3 we get that the following systems satisfy the matrix condition

(6.54) $$(z_{1,2}, z_{1,3}; z_{3,4}, -z_{2,4})$$

(6.55) $$(z_{1,3}, z_{1,4}; z_{2,4}, -z_{2,3})$$

(6.56) $$(z_{1,2}, z_{1,4}; z_{3,4}, z_{2,3})$$

Using these facts for all possible $1 \leq i_1 < i_2 < i_3 < i_4 \leq n$, and using of course the fact that $5 \leq n$, we get that if $1 \leq i < j < k \leq n$ then

(6.57) $$y_{i,k} G y_{j,k}, \quad y_{i,j} G y_{j,k}, \quad y_{i,j} G y_{i,k}$$

and if $1 \leq i_1 < j_1 \leq n$, $1 \leq i_2 < j_2 \leq n$ and $\{i_1,j_1\} \cap \{i_2,j_2\} = \emptyset$ then

(6.58) $$y_{i_1,j_1} \perp y_{i_2,j_2}$$

Now (6.57) implies in particular that $y_{1,2} G y_{1,3}$ $y_{1,2} G y_{1,4}$ $y_{1,2} G y_{1,5}$ and $y_{1,2} G y_{2,5}$. So

(6.59) $$y_{1,2} = b_{1,2} + c_{1,2}$$

(6.60) $$b_{1,2} = \ell(b) y_{1,2}\, \ell(w_2); \qquad c_{1,2} = \ell(u_2) y_{1,2}\, r(c)$$

If $3 \leq i < j \leq n$, we get in a similar way (using again $5 \leq n$) that

(6.61) $$y_{i,j} = b_{i,j} + c_{i,j}$$

(6.62) $$b_{i,j} = \ell(\tilde{u}_i) y_{i,j}\, \ell(\tilde{w}_j); \qquad c_{i,j} = \ell(\tilde{u}_j) y_{i,j}\, \ell(\tilde{w}_i)$$

Also, if $3 \leq i < j \leq n$ then $(y_{1,2}/2, y_{1,i}/2; y_{i,j}/2, -y_{2,j}/2)$ satisfies M (since if we define $i_1 = 1$, $i_2 = 2$, $i_3 = i$ and $i_4 = j$, then this system is the system in (6.54)). So

(6.63) $|b_{1,2}| = |c_{2,j}|$, $|c_{1,2}| = |c_{1,i}|$, $|b_{1,i}| = |c_{i,j}|$, $|b_{2,j}| = |b_{i,j}|$

(6.64) $|b^*_{1,2}| = |b^*_{1,i}|$, $|c^*_{1,2}| = |b^*_{2,j}|$, $|c^*_{1,i}| = |b^*_{i,j}|$, $|c^*_{2,j}| = |c^*_{i,j}|$

(6.65) $v(b_{1,i})^* v(b_{1,2}) = v(c_{i,j})^* v(c_{2,j})$; $v(c_{2,j}) v(b_{1,2})^* = v(c_{i,j}) v(b_{1,i})^*$

(6.66) $v(b_{2,j})^* v(c_{1,2}) = v(b_{i,j})^* v(c_{1,i})$; $v(c_{1,i}) v(c_{1,2})^* = v(b_{i,j}) v(b_{2,j})^*$

Note that this implies in particular that $\|b\|_1 = \|c\|_1 = 1$. Also the system $(y_{1,2}/2, y_{1,j}/2; y_{i,j}/2, y_{2,i}/2)$ satisfies M (since, if we use again $i_1 = 1$, $i_2 = 2$, $i_3 = i$ and $i_4 = j$, then this system is just the system in (6.56)). So:

(6.67) $v(b_{1,j})^* v(b_{1,2}) = -v(b_{i,j})^* v(c_{2,i})$; $v(b_{2,i})^* v(c_{1,2}) = -v(c_{i,j}) v(c_{1,j})$

(6.68) $v(c_{2,i}) v(b_{1,2})^* = -v(b_{i,j}) v(b_{1,j})^*$; $v(c_{1,j}) v(c_{1,2})^* = -v(c_{i,j}) v(b_{2,i})$

Hence,

$$(6.69)\quad c_{1,2} = v(c_{1,2}) v(c_{1,i})^* c_{1,i} = v(c_{1,2}) v(c_{1,i})^* c_{2,i} w_2$$

$$= (v(c_{1,2}) v(c_{1,i})^*) \cdot (v(c_{2,i}) v(b_{1,2})^*) b_{1,2} w_2$$

$$= (v(b_{2,j}) v(b_{i,j})^*) \cdot (-v(b_{i,j}) v(b_{1,j})^*) b_{1,2} w_2$$

$$= -u_2 b_{1,2} w_2$$

Put

(6.70) $$a = b_{1,2} w_2$$

and for $3 \leq j \leq n$:

(6.71) $$w_j = v(b_{1,j})^* v(b_{1,2}) w_2; \quad u_j = v(c_{1,j}) v(c_{1,2})^* u_2$$

Then, for every $1 \leq j \leq n$ we get

(6.72) $$v(w_j) = r(a); \quad r(u_j) = \ell(a)$$

(6.73) $$\ell(w_{j_1}) \cdot \ell(w_{j_2}) = 0 = \ell(u_{j_1}) \cdot \ell(u_{j_2}), \; 1 \leq j_1 \neq j_2 \leq n$$

We shall show that for every $1 \leq i < j \leq n$

(6.74) $$c_{i,j} = -u_j a w_i^*, \quad b_{i,j} = u_i a w_j^*$$

If $i = 1$, $j = 2$, then (6.74) follows from (6.69) and (6.70). If $3 \leq j \leq n$ and $i = 1$, then from (6.63), (6.64), (6.69), (6.70) and (6.71) we get

(6.75) $$b_{1,j} = |b_{1,j}^*| v(b_{1,j}) = |b_{1,2}^*| v(b_{1,j}) = b_{1,2} v(b_{1,2})^* v(b_{1,j})$$

$$= a w_2^* v(b_{1,2})^* v(b_{1,j}) = a w_j^* = u_1 a w_j^*$$

(6.76) $$c_{1,j} = v(c_{1,j}) |c_{1,j}| = v(c_{1,j}) |c_{1,2}| = v(c_{1,j}) v(c_{1,2})^* c_{1,2}$$

$$= -v(c_{1,j}) v(c_{1,2})^* u_2 a = -u_j a = -u_j a w_1^*$$

If $3 \leq j \leq n$ and $i = 2$, then we get (6.74) from (6.75), (6.76) and (6.52).

Finally, if $3 \leq i < j \leq n$, then using (6.63), (6.66), (6.68), (6.71) we get:

$$(6.77) \qquad b_{i,j} = v(b_{i,j})v(b_{2,j})^* b_{2,j} = v(c_{1,i})v(c_{1,2})^* u_2 a w_j^* = u_i a w_j^*$$

$$(6.78) \qquad c_{i,j} = v(c_{i,j})v(b_{2,i})^* b_{2,i} = -v(c_{1,j})v(c_{1,2})^* u_2 a w_i^* = -u_j a w_i^*$$

So, (6.74) is valid for every $1 \leq i < j \leq n$. This formula can be written in an appropriate tensor-product representation as

$$(6.79) \qquad c_{i,j} = -e_{j,i} \otimes a; \quad b_{i,j} = e_{i,j} \otimes a$$

so

$$(6.80) \qquad y_{i,j} = b_{i,j} + c_{i,j} = (e_{i,j} - e_{j,i}) \otimes a = a_{i,j} \otimes a$$

where $a_{i,j} = e_{i,j} - e_{j,i}$, $1 \leq i < j \leq n$, are the elementary anti-symmetric matrices. So $V(X) = \overline{\mathrm{span}}\{y_{i,j}\}_{1\leq i<j\leq n} = A_1^n(a)$ and thus $V(X)$ is an elementary subspace of C_1 of type 3. This completes the proof in case X is of type 3. Thus, if X is an elementary subspace of C_1 of type different from 4, and V is an isometry from X into C_1 then $V(X)$ is an elementary subspace of C_1 (of the same type as X). By the results of §2 there exists a contractive projection from C_1 onto any elementary subspace and in particular onto $V(X)$. □

Remark 6.5 Theorem 6.1 does not hold if X is of type 4, i.e. there exist subspaces of C_1 which are isometric to a Hilbert space which are not elementary subspaces of C_1. Consider the following simple example.

Let $0 \leq \alpha,\beta \leq 1$, $\alpha+\beta = 1$ and let

$$(6.18)\qquad x = \begin{pmatrix} 1 & 0 \\ 0 & 0 \end{pmatrix} \; ; \qquad y = \begin{pmatrix} 0 & \alpha \\ \beta & 0 \end{pmatrix}$$

Using the following formula to compute the C_1-norm of a 2×2 matrix A:

$$(6.82)\qquad \|A\|_1 = (\mathrm{trace}(A^*A) + 2|\det A|)^{1/2}$$

we get easily that for every choice of scalars s,t:

$$(6.38)\qquad \|sx + ty\|_1 = (|s|^2 + |t|^2)^{1/2}$$

So, $X = \mathrm{span}\{x,y\}$ is isometric to ℓ_2^2. Nevertheless, X is an elementary subspace of C_1 if and only if $\alpha.\beta = 0$ (in an elementary Hilbert subspace all the non zero elements have the same rank). Taking α,β so that $\alpha\cdot\beta \neq 0$ we get an example of non-elementary Hilbert subspace of C_1.

If X is an elementary Hilbert subspace of C_1, and V is an isometry from X into C_1 which preserves not only the norm but also the G-relation, then $V(X)$ is also an elementary subspace of C_1. This follows from Theorem 6.2.

Proof of Theorem 6.2 Let H be a Hilbert space and let V be an isometry from H into C_1, which satisfies (6.1). Clearly, H must be separable, so let $\{e_k\}_{k=1}^n$, $1 \leq n \leq \infty$, be an orthonormal basis of H. Put $z_k = Ve_k$, $1 \leq k \leq n$. Using the condition (6.1) we get that $\{z_k\}_{k=1}^n$ satisfies SG (see Definition 5.13). By the arguments presented in §5e, as summarized in Corollary 5.18, we get that $V(H) = \overline{\mathrm{span}}\{z_k\}_{k=1}^n$ is an elmentary Hilbert subspace of C_1. In particular, there exists a contractive projection from C_1 onto $V(H)$. □

We conclude this section with some simple corollaries and remarks.

Corollary 6.6 Let V be an isometry from SY_1 onto itself, or from A_1 onto itself. Then there exists a unitary operator w on ℓ_2 so that for every x

(6.84) $Vx = w^T x w$

(where w^T is the transpose of w in the given matrix representation).

Proof We prove the corollary in case V is an isometry from A_1 onto itself; the case of SY_1 is proved similarly. As in Theorem 6.1 we get that for every $1 \leq i < j < \infty$:

(6.85) $Va_{i,j} = u_i a w_j^* - u_j a w_i^*$

for some $a \in C_1$, $\|a\|_1$, and for some partial isometries $\{u_i\}_{i=1}^\infty$ and $\{w_i\}_{i=1}^\infty$ with $r(u_i) = \ell(a) = u_1$, $r(w_i) = r(a) = w_1$ and $\ell(u_i)\ell(u_j) = 0 = \ell(w_i)\ell(w_j)$ for $i \neq j$.

Since V is onto A_1 we get $\sum_{i=1}^\infty \ell(u_i) = 1 = \sum_{i=1}^\infty \ell(w_i)$ (the series coverage in the weak operator topology). If $0 \neq x$ is an anti-symmetric matrix it is easy to verify that $\text{rank}(x) \geq 2$. Thus from the formula $Vx = x \otimes a$, which expresses (6.85) in an appropriate tensor-product representation, we get

(6.86) $\text{rank}(Vx) = \text{rank}(x \otimes a) = \text{rank}(x)\cdot\text{rank}(a) \geq 2\ \text{rank}(a)$

Since V is onto A_1, there exists $x \in A_1$ with $\text{rank}(Vx) = 2$, and this implies by (6.86) that $\text{rank}(a) = 1$. This implies the existence of unitary operators $\tilde{w}$ and $\tilde{u}$ on ℓ_2, so that

(6.87) $Vx = \tilde{u} x \tilde{w}$, $x \in A_1$

Since $Vx \in A_1$, we get from this

(6.88) $\tilde{u}x\tilde{w} = \tilde{w}^T x \tilde{u}^T, \quad x \in A_1$

From this we get easily that $\tilde{u} = \lambda\tilde{w}^T$, $|\lambda| = 1$. Let α be any square root of λ and put $w = \alpha\tilde{w}$. Then (6.84) follows from (6.87) and $\tilde{u} = \lambda\tilde{w}^T$. □

Corollary 6.7 In the following list of subspaces of C_1 there exist no mutually isometric spaces.

(6.89) $C_1^{n,m}, (2 \leq n \leq m \leq \infty,\ 3 \leq m)$; $SY_1^n, (1 \leq n \leq \infty)$; $A_1^n, (5 \leq n \leq \infty)$;

$\ell_2^n, (2 \leq n \leq \infty)$; $AH_1^n, (2 \leq n < \infty)$; $DAH_1^n, (3 \leq n < \infty)$.

Proof: It is clear that $\ell_2^n, (2 \leq n \leq \infty)$, is not isometric to any other space in the list, since each of them (except SY_1^1 which is one-dimensional) contains subspace isometric to ℓ_1^2 (two dimensional ℓ_1-space). The dimension of AH_1^n is $2n$ and the dimension of DAH_1^n is $2n-1$; so AH_1^n is not isometric to DAH_1^m for any n and m. Let us denote by $s(X)$ the maximal s so that ℓ_1^s is isometric to a subspace of X. Then it is clear that

(6.90) $s(C_1^{n,m}) = \min\{n,m\}$; $s(SY_1^n) = n$; $s(A_1^n) = [\frac{n}{2}]$

$s(AH_1^n) = s(DAH_1^n) = 2.$

Since an isometry preserves the dimension and the number "$s(X)$", it is clear from (6.90) and in some cases by Theorem 6.1 that no two finite dimensional spaces from the list (6.89) are isometric.

Also, it is clear that no pair of spaces from the sequence $\{C_1^{n,\infty}\}_{n=2}^{\infty}$ are isometric, and that by Theorem 6.1 the spaces A_1 and SY_1 are not isometric to any one of the spaces $C_1^{n,\infty}$ or C_1.

It remains to show that SY_1 is not isometric to A_1. Assume the contrary, and let V be an isometry from SY_1 onto A_1. By Theorem 6.1 there exists a tensor-product representation and an element $a \in C_1$ so that $Vx = x \otimes a$ for any $x \in SY_1$. As in the proof of Corollary 6.6 we must have $\mathrm{rank}(a) \geq 2$. Using the isometry V^{-1} from A_1 onto SY_1 we get by Theorem 6.1 and similar arguments that for any $y \in A_1$, $\mathrm{rank}(V^{-1}y) \geq \mathrm{rank}(y)$. If $x = e_{1,1} \in SY_1$, then

$$(6.91) \qquad 1 = \mathrm{rank}(x) = \mathrm{rank}(V^{-1}(Vx)) \geq \mathrm{rank}(Vx) = \mathrm{rank}(x \otimes a)$$

$$= \mathrm{rank}(x)\cdot\mathrm{rank}(a) \geq 2$$

This contradiction shows that SY_1 is not isometric to A_1. □

Let us turn to the uniqueness problems posed in the end of §2. Let X be the range of a contractive projection in C_1. By Theorem 2.14, $X = \sum_{k\in K} X_k$, $1 \leq |K| \leq \aleph_0$, where each X_k is an elementary subspace of C_1, which is a subspace of X, and $X_k \perp X_{k'}$ for every $k \neq k'$. In particular, X is isometric to the ℓ_1-sum of the spaces $\{X_k\}_{k\in K}$, and thus to a space of the form $Y = (\sum_{k\in K} \oplus Y_k)_{\ell_1}$ where each Y_k is a <u>simple</u> elementary subspace of C_1 isometric to X_k (the simple elementary subspaces of C_1 are the spaces in the list (6.89)).

<u>Corollary 6.8</u> (i) the representation $X = \sum_{k\in K} X_k$ of a subspace X of C_1 which is the range of a contractive projection from C_1 as a sum of pairwise disjointly supported elementary subspaces - is unique.

(ii) Let $\{Y_k\}_{k\in K}$, $1 \leq |K| \leq \aleph_0$, be a family of simple elementary subspaces of C_1 and let V be an isometry from $Y = (\sum_{k\in K} \oplus Y_k)_{\ell_1}$ into C_1. Assume also that $V|_{Y_k}$ preserves the G-relation if Y_k is some ℓ_2^n. Let $V(Y_k) = X_k$ and $V(Y) = X$. Then $X = \sum_{k\in K} X_k$, the $\{X_k\}_{k\in K}$ are disjointly supported, and there is a contractive projection from C_1 onto X.

Proof: (i) By Proposition 5.2 and the proof of Theorem 2.14 (given in §5g), the representation $X = \sum_{k \in K} X_k$ is just the prepresentation of X as the sum of the family of its irreducible components, which are pairwise disjointly supported. This representation is uniquely determined by X.

(ii) By Theorem 6.1 and 6.2, $X_k = V(Y_k)$ is an elementary subspace of C_1 (of the same type as Y_k). If $k \neq k'$ and $x \in Y_k$, $y \in Y_{k'}$ are normalized elements, then for every scalar s,t

$$(6.9) \quad \|sVx+tVy\|_1 = \|sx+ty\|_Y = |s|+|t|$$

so by Proposition 1.1 we have $Vx \perp Vy$. Thus $X_k \perp X_{k'}$ for every $k,k' \in K$ with $k \neq k'$. If Q_k is the canonical contractive projection from C_1 onto X_k (so, $Q_k = Q_k E(X_k)$) then $Q = \sum_{k \in K} Q_k$ is a contractive projection from C_1 onto X. □

Remark 6.9 C_1 is isometric to a subspace of A_1 and to a subspace of SY_1.

Proof Define mappings $V: C_1 \to A_1$ and $W: C_1 \to SY_1$ by

$$(6.93) \quad Ve_{i,j} = \begin{cases} \frac{1}{2}a_{2i,2j-1} \ ; & i < j \\ \frac{1}{2}a_{2j-1,2i} \ ; & i \geq j \end{cases}$$

and

$$(6.94) \quad We_{i,j} = \begin{cases} \frac{1}{2}s_{2i,2j-1} \ ; & i < j \\ \frac{1}{2}s_{2j-1,2i} \ ; & i \geq j \end{cases}$$

An easy computation gives that V and W are isometries into. □

Up to now we dealt in this section with isometries of elementary subspaces of C_1 into C_1. We do not know if the analogous results hold for isometries of elementary subspaces of C_∞ into C_∞. The key to this problem is the following question: "is the matrix or the diagonal conditions invariant under isometries into C_∞?"

Our guess is the following:

<u>Conjecture 6.10</u> Let $x_1, x_2, \tilde{x}_1, \tilde{x}_2$ be normalized elements of C_∞ so that for every choice of scalars $t_1, t_2, \tilde{t}_1, \tilde{t}_2$ we have

$$(6.95)\qquad \|t_1x_1+t_2x_2+\tilde{t}_1\tilde{x}_1+\tilde{t}_2\tilde{x}_2\|_\infty = \left\|\begin{pmatrix} t_1 & t_2 \\ -\tilde{t}_2 & \tilde{t}_1 \end{pmatrix}\right\|_\infty$$

Then,

$$(6.96)\qquad x_k = y_k+z_k; \quad \tilde{x}_k = \tilde{y}_k+\tilde{z}_k; \quad \|y_k\|_\infty = \|\tilde{y}_k\|_\infty = 1$$

where the system $(y_1, y_2; \tilde{y}_1, \tilde{y}_2)$ satisfies the matrix condition, $\mathrm{span}\{y_j, \tilde{y}_j\}_{j=1}^2 \perp \mathrm{span}\{z_j, \tilde{z}_j\}_{j=1}^2$, and for every scalar $t_1, t_2, \tilde{t}_1, \tilde{t}_2$ we have

$$(6.97)\qquad \|t_1z_1+t_2z_2+\tilde{t}_1\tilde{z}_1+\tilde{t}_2\tilde{z}_2\|_\infty \leq \left\|\begin{pmatrix} t_1 & t_2 \\ -\tilde{t}_2 & \tilde{t}_1 \end{pmatrix}\right\|_\infty$$

<u>Conjecture 6.11</u> Let $x, \tilde{x}, y$ be normalized elements in C_∞ so that for every scalar s, $\tilde{s}$ and t:

$$(6.98)\qquad \|sx+\tilde{s}\tilde{x}+ty\|_\infty = \left\|\begin{pmatrix} s & t/2 \\ t/2 & -\tilde{s} \end{pmatrix}\right\|_\infty$$

Then,

(6.99) $x = x_1+x_2, \quad \tilde{x} = \tilde{x}_1+\tilde{x}_2, \quad y = y_1+y_2$

where the normalized system $(x_1,\tilde{x}_1,y_1)$ satisfies the diagonal condition, $\text{span}\{x_1,\tilde{x}_1,y_1\} \perp \text{span}\{x_2,\tilde{x}_2,y_2\}$, and for every scalar s, $\tilde{s}$ and t we have:

(6.100) $$\|sx_2+\tilde{s}\tilde{x}_2+ty_2\|_\infty \leq \left\|\begin{pmatrix} s & t/2 \\ t/2 & -\tilde{s} \end{pmatrix}\right\|_\infty .$$

§7. APPLICATIONS

The information we obtianed in the previous sections on a single contractive projection in C_1 or in C_∞ can be used in solving problems concerning families of contractive projections in these spaces.

In [4], Gleit and Lazar conjectured that the basis constant of C_1 and C_∞ is strictly greater than one. We shall prove below that these spaces do not have monotone bases, and thus partially substantiate this conjecture. Our approach in proving the inexistence of a monotone basis is to assume the converse, namely the existence of a sequence $\{P_n\}_{n=1}^{\infty}$ of contractive projections with $P_nP_m = P_{\min\{n,m\}}$, $\dim R(P_n - P_{n-1}) = 1$, and $P_n \underset{n\to\infty}{\longrightarrow} I$ is the strong operator topology, and then to arrive at a contradiction.

We first establish some facts concerning sequences of contractive projections in C_1 with one dimensional gaps. If P is a contractive projection in C_1 and if x is an atom of P then the irreducible component of $R(P)$ which contains x is denoted by $C(x,P)$. By the analysis made in §5, $C(x,P)$ is an elementary subspace of C_1.

<u>Proposition 7.1</u> Let P and Q be contractive projections in C_1 with $PQ = QP = P$ and $\dim R(Q-P) = 1$. Let x and y be atoms of P with $x\dot{G}y$. Then x and y are also atoms of Q and

(7.1) $R(F(x)G(y)P) = R(F(x)G(y)Q)$

(7.2) $R(F(y)G(x)P) = R(F(y)G(x)Q)$

In particular if $C(x,P)$ is of type 4 (i.e. $C(x,P)$ is an elementary Hilbert subspace of C_1) then $C(x,Q)$ is also of type 4.

Proof: Put $X = R(E(x)Q)$. If $\dim X = 2$, then by Theorem 2.14 X is either an elementary Hilbert subspace of C_1, or $X = \mathrm{span}\{x_1,x_2\}$ where $x_1 \perp x_2$ and $x = x_1+x_2$. Now the first alternative cannot happen since then if we choose $x' \in X$ $\|x'\| = 1$ so that $\{x,x'\}$ is an orthonormal basis of X then xGx', and this contradicts the fact that $E(x) = E(X)$. Also the second alternative cannot happen. Indeed put $y_j = G(x_j)y$ $j = 1,2$. Then $0 \neq y_j \in R(Q)$, $y = y_1+y_2$; and since x and y are atoms of P (and thus $\dim R(E(x)P) = \dim R(E(y)P) = 1$), we get:

$$(7.3) \quad 1 = \dim R(Q-P) \geq \dim R(E(x)(Q-P)) + \dim R(G(x)(Q-P)) \geq 1+1 = 2,$$

a contradiction.

It follows that $\dim X = 1$, and thus x is an atom of Q. By symmetry, y is also an atom of Q.

Let us prove (7.1). Since $Q \geq P$, we clearly have $R(F(x)G(y)P) \subseteq R(F(x)G(y)Q)$. If $\tilde{x}$ is an atom of $F(x)G(y)Q$, then by Proposition 4.6 there exists an atom $\tilde{y}$ of $F(y)G(x)Q$ so that the system $(x,y;\tilde{x},\tilde{y})$ satisfies the matrix condition. By Proposition 3.6, $\tilde{x} \in R(P)$ if and only if $\tilde{y} \in R(P)$. Since $\dim R(Q-P) = 1$ we must have that $\tilde{x} \in R(P)$ and $\tilde{y} \in R(P)$. This clearly establishes (7.1). By symmetry we have also (7.2).

Recall that $C(x,P)$ is of type 4 if and only if $F(x)G(y)P = 0$. By (7.1) we get $F(x)G(y)Q = 0$ and since x,y are atoms of Q this is equivalent to the fact that $C(x,Q)$ is of type 4. □

Proposition 7.2 Let P and Q be contractive projections in C_1 so that $PQ = QP = P$ and $\dim R(Q-P) = 1$. Let x and $\tilde{x}$ be atoms of P, and let $y \in R(P)$ so that the system $(x,\tilde{x},y)$ satisfies the diagonal condition. Then x and $\tilde{x}$ are atoms of Q.

Proof: Assume that $\dim R(E(x)Q) = 2$. As in the proof of Proposition 7.1, $R(E(x)Q) = \text{span}\{x_1,x_2\}$ where $x_1 \perp x_2$ and $x = x_1+x_2$. Put $y_j = G(x_j)y$, $j = 1,2$. Then $0 \neq y_j \in R(Q)$ and $y = y_1+y_2$. Put also $\tilde{x}_1 = F(y_2)E(y_1)\tilde{x}$, $\tilde{x}_2 = F(y_1)E(y_2)\tilde{x}$. Then $0 \neq \tilde{x}_j \in R(Q)$, $\tilde{x}_1 \perp \tilde{x}_2$ and $\tilde{x} = \tilde{x}_1+\tilde{x}_2$. Since x and y are atoms of P, we get

$$(7.4) \quad 1 = \dim R(Q-P) \geq \dim R(E(x)(Q-P)) + \dim R(G(x)(Q-P)) \geq 1+1 = 2 \;,$$

a contradiction. Thus $\dim R(E(x)Q) = 1$ and x is an atom of Q. By symmetry, also $\tilde{x}$ is an atom of Q.

Proposition 7.3 Let $\{P_k\}_{k=1}^n$ be a finite sequence of contractive projections in C_1 so that $P_kP_\ell = P_{\min\{k,\ell\}}$ and $\dim R(P_{k+1}-P_k) = 1$ for $1 \leq k \leq n$.

(i) Let x be an atom of P_1 so that $C(x,P_1)$ is of type 4. Then $C(x,P_n)$ is also of type 4.

(ii) Let x be an atom of P_1 so that $C(x,P_1) = \text{span}\{x,\tilde{x},y\}$ where $(x,\tilde{x},y)$ is a normalized system which satisfies the diagonal condition. Then x and $\tilde{x}$ are atoms of P_n, $G(y)P_n = 0$, and if $C(x,P_n) \neq C(x,P_1)$ then $C(x,P_n)$ is of type 5 or 6.

Proof (i) Being an elementary subspace of type 4, $C(x,P_1)$ has by definition dimension at least two. Using Proposition 7.1 we get by a simple induction argument for every $1 \leq k \leq n$ $C(x,P_k)$ is also of type 4.

(ii) Using Proposition 7.2 we get by induction that x and $\tilde{x}$ are atoms of P_k for every $1 \leq k \leq n$. Let $1 \leq m \leq n$ be the maximal index so that

$$(7.5) \qquad R(E(y)P_k) = R(P_1) = \text{span}\{x,\tilde{x},y\}, \qquad 1 \leq k \leq m \;.$$

We claim that for every $1 \leq k \leq m$, $G(y)P_k = 0$, and thus $C(x,P_k) = \text{span}\{x,\tilde{x},y\}$. We prove this by induction and thus it is sufficient to prove this for $k = 2$ (in case $m > 1$). If $G(y)P_2 \neq 0$, then by Proposition 5.3 and §5(b) $C(x,P_2)$ is an elementary subspace of type two, i.e. $C(x,P_2) = SY_1^s(x)$, $3 \leq s$. In particular, $\dim R(G(y)P_2) = \dim R(G(y)(P_2-P_1)) \geq 2$, which contradicts the assumption $\dim R(P_2-P_1) = 1$. So, indeed, $G(y)P_k = 0$ for every $1 \leq k \leq m$.

If $m = n$, we are done. If $m < n$ then clearly there exist atoms z and $\tilde{z}$ of P_{m+1} so that the system $(x,z;\tilde{x},\tilde{z})$ satisfies the matrix condition, $y = (z+\tilde{z})/2$ and $R(E(y)P_{m+1}) = \text{span}\{x,z,\tilde{x},\tilde{z}\}$. Since $\dim R(P_{m+1}-P_m) = 1$, we clearly have $G(y)P_{m+1} = 0$. Now, the system $(z,\tilde{z},(x+\tilde{x})/2)$ satisfies the diagonal condition, and thus using Proposition 7.2 we get by induction that z and $\tilde{z}$ are atoms of P_k for every $m+1 \leq k \leq n$. In order to show that for $m+1 \leq k \leq n$ $C(x,P_k)$ is of type 5 or 6 it is enough by Proposition 5.3 and the proof of Theorem 2.14 to show that $G(y)P_k = 0$, and as before it is enough to prove this to $k = m+2$. If $G(y)P_{m+2} \neq 0$, then by the analysis made before Lemma 4.9, we have

$$(7.6) \qquad 2 \leq \dim R(G(y)P_{m+2}) = \dim R(G(y)(P_{m+2}-P_{m+1}))$$

$$\leq \dim R(P_{m+2}-P_{m+1}) = 1$$

a contradiction. So, indeed $G(y)P_k = 0$ for every $m+1 \leq k \leq n$ and $C(x,P_k)$ is of type 5 or 6. □

<u>Corollary 7.4</u> Let $\{P_k\}_{k=1}^n$ be as in Proposition 7.3, and assume in addition that $\dim R(P_1) = 1$. Then for every atom x of P_n, $C(x,P_n)$ is of type 2, 4, 5 or 6. In case $C(x,P_n)$ is of type 2 it is either of the form $SY_1^2(x)$, or it is one dimensional.

We are ready now to consider the problem of monotone bases in C_1 and C_∞. Our negative result is

Proposition 7.5 There is no monotone w^*-basis in C_1. Thus the spaces C_1 and C_∞ do not have monotone bases.

Proof It is clear that the first statement implies the second. If there exists a w^*-monotone basis in C_1, then there exists a sequence $\{P_k\}_{k=1}^\infty$ of contractive projections in C_1 with $P_k P_\ell = P_{\min\{k,\ell\}}$, $\dim R(P_k - P_{k-1}) = 1$ for $1 \leq k < \infty$ (where $P_0 = 0$), and for every $x \in C_1$ and $y \in C_\infty$ we have $(P_k x, y) \underset{k\to\infty}{\longrightarrow} (x,y)$.

Let n be the first index such that $R(P_n)$ has less than n irreducible components. Let $\{x_j\}_{j=1}^{n-1}$ be atoms of P_{n-1} with pairwise disjoint supports and let $x \in R(P_n)$, $x \notin R(P_{n-1})$, $\|x\|_1 = 1$. Now,

$$(7.7) \qquad x = \Big(\prod_{j=1}^{n-1} F(x_j)\Big)x + \sum_{j=1}^{n-1} E(x_j)x + \sum_{1\leq i<j\leq n-1} G(x_j)G(x_i)x$$

$$+ \sum_{j=1}^{n-1} G(x_j)\Big(\prod_{\substack{i=1\\ i\neq j}}^{n-1} F(x_i)\Big)x$$

All the summands in (7.7) belong to $R(P_n)$, and since $\dim R(P_n - P_{n-1}) = 1$, only one of them does not vanish. Clearly $\prod_{j=1}^{n-1} F(x_j)x = 0$, since otherwise we get a contradiction to the choice of n. If $E(x_j)x \neq 0$ for some $1 \leq j \leq n-1$, then as in the proof of Proposition 7.1, $R(E(x_j)P_n)$ is generated by two elements with disjoint supports and thus $R(P_n)$ has n components, a contradiction.

It follows that one (and only one) of the following possibilities occurs:

(a) for some j, $x = G(x_j)x$ and $F(x_i)x = 0$ for every $1 \leq i \leq n-1$ with $i \neq j$;

(b) for some $i < j$, $x = G(x_i)G(x_j)x$ and $F(x_k)x = 0$ for every $1 \leq k \leq n-1$ with $j \neq k \neq i$.

In case (a), $C(x,P_n)$ is of type 4, and thus x is an atom of P_m for every $m \geq n$, and $C(x,P_m)$ is of type 4. Since $(P_m z,y) \underset{m\to\infty}{\longrightarrow} (z,y)$ for every $z \in C_1$ and $y \in C_\infty$, we get that x must be of rank 1. Since $x_j G x$, $\operatorname{rank}(x_j) = 1$, and thus $F(x)G(x_j) \neq 0$ but $F(x)G(x_j)P_m = 0$ for every $n \leq m$. For every $z \in C_1$ and $y \in C_\infty$ we get

$$(7.8) \qquad 0 = \lim_{m\to\infty}(F(x)G(x_j)P_n z,y) = \lim_{m\to\infty}(P_m z,F(x)G(x_j)y)$$

$$= (z,F(x)G(x_j)y)$$

So, $F(x)G(x_j) = 0$, a contradiction.

In case (b) we can choose x so that the normalized system (x_i,x_j,x) satisfies the diagonal condition. Clearly, $C(x_i,P_n) = \operatorname{span}\{x_i,x_j,x\}$. By Proposition 7.3, x_i and x_j are atoms of P_m for every $m \geq n$ and $G(x)P_m = 0$.

Again, the atomicity of x_i and x_j with respect to P_m for every $m \geq n$ implies that $\operatorname{rank}(x_i) = \operatorname{rank}(x_j) = 1$, so $\operatorname{rank}(x) = 2$. Since $G(x) \neq 0$ and $G(x)P_m = 0$ for every $m \geq n$, we get a contradiction as in case (a). □

Remarks 7.6 (i) The proof we gave to Proposition 7.5 works also in $C_1^{n,m}$, $2 \leq n \leq m \leq \infty$, $3 \leq m$. Thus an elementary subspace of C_1 of type 1 does not have a monotone w^*-basis, and an elementary subspace of C_∞ of type 1 does not have a monotone basis.

(ii) Let $\{e_i\}_{i=1}^{\infty}$ be any orthonormal basis of ℓ_2, let $a_n = \sum_{i=1}^{n} (\cdot e_i)e_i$, and define $P_n x = a_n x a_n$. Then P_n is a contractive projection in both C_1 and C_∞, and $P_n \to I$ in the strong operator topology. Thus, C_1 and C_∞ admit a finite dimensional monotone Schauder decomposition. Note that $\dim R(P_n - P_{n-1}) = 2n-1$. A natural question is therefore whether this can be improved:

(*) Does there exist a sequence $\{P_n\}_{n=1}^{\infty}$ of contractive projections in either C_1 or C_∞ so that $P_n P_m = P_{\min\{n,m\}}$, $P_n x \underset{n\to\infty}{\longrightarrow} x$ for every x, and

(7.9) $\sup_n \dim R(P_n - P_{n-1}) < \infty$

(iii) Let $e_{i,j}$ be the standard unit matrices, and arrange them in one sequence in the following way:

$$e_{1,1}, e_{1,2}, e_{2,2}, e_{2,1}, e_{1,3}, e_{2,3}, e_{3,3}, e_{3,1}, e_{3,2}, \dots$$

(7.10)

$$\dots, e_{1,n}, e_{2,n}, \dots, e_{n,n}, e_{n,1}, e_{n,2}, \dots, e_{n,n-1}, \dots$$

Then it can be easily verified that the sequence (7.10) is a Schauder basis in C_1 and in C_∞, and its basis constant is $\leqslant 2$.

Another simple application is the description of all possible contractive projections whose complement is also contractive.

Proposition 7.7 Let P be a contractive projection in C_1 so that $Q = I-P$ is also contractive. Then one of the following mutually exclusive cases occurs:

(i) There exist projections a, b in ℓ_2 so that

(7.11) $Px = axb + (I-a)x(I-b)$

(7.12) $Qx = ax(I-b) + (I-a)xb$

(ii) There exists an isometry of C_1 of the form $Vx = uxw$ with u and w unitary operators on ℓ_2, so that either

(7.13) $VPV^{-1}x = \frac{x+x^T}{2}$, $VQV^{-1}x = \frac{x-x^T}{2}$

or

$$(7.14)\quad VPV^{-1}x = \frac{x-x^T}{2}, \quad VQV^{-1}x = \frac{x+x^T}{2}$$

(where x^T is the transpose of x relative to some given matrix representation).

Proof: Put $X = R(P)$ and $Y = R(Q)$. Using Proposition 3.6 it is very easy to show that if a_1, a_2, b_1, b_2 are non zero projections in ℓ_2 so that $a_1a_2 = 0$, $b_1b_2 = 0$ and

$$(7.15)\quad a_1C_1b_1 + a_1C_1b_2 + a_2C_1b_1 \subseteq X$$

then also $a_2C_1b_2 \subseteq X$.

From this it follows easily that X and Y have at most two non-trivial irreducible components, every atom of either P or Q has rank ≤ 2, and that $F(X) = F(Y) = 0$.

Case 1: X has only one irreducible component and $G(X) = 0$. In this case X is an elementary subspace, $E(X) = I_{C_1}$ and P is the canonical contractive projection from C_1 onto X in the sense of §2. A brief check yields that either X is of type 2 and Y is of type 3, or conversely. That is, with respect to some matrix representation we have either $X = SY_1$ and $Y = A_1$, or $X = A_1$ and $Y = SY_1$, and in both cases P and Q are the canonical projections onto X and Y. This is clearly equivalent to the existence of unitary operators u and w on ℓ_2 so that in the first case we have (7.13) and in the second case we have (7.14).

Case 2: X has two irreducible components, or X has only one irreducible component but $G(X) \neq 0$.

Assume first that X has two irreducible components X_1 and X_2. Using the observation in the beginning of the proof, we get that $G(X) = 0$. Let $Y_1 = \ell(X_1)C_1r(X_2)$, $Y_2 = \ell(X_2)C_1r(X_1)$. Then Y_1 and Y_2 are subspaces of Y

and thus $G(Y) = 0$. Y has two irreducible components, for if not then using case 1 we get a contradiction to the assumption that X has two irreducible components. It follows that $QE(X_1) = QE(X_2) = 0$ and $Y = Y_1+Y_2$. Since $E(X) = E(Y) = I_{C_1}$, P and Q are the canonical projections from C_1 onto X and Y, and thus we have (7.11) and (7.12) with $a = \ell(X_1)$ and $b = r(X_1)$.

Next, assume that X and Y are irreducible and that $G(X) \neq 0 \neq G(Y)$. Since $F(X) = F(Y) = 0$ we can assume that $\ell(X) = \ell(Y) = I$ and that $r(X) < I$, $r(Y) < I$. Put $Q_1 = QE(X)$, then $Q = E(X)Q$ and thus Q_1 is a projection. If $Q_1 \neq 0$, let y be an atom of Q_1 and put $a_1 = \ell(y)$, $a_2 = I-a_1$, $b_1 = I-r(X)$ and $b_2 = r(y)$. Clearly, $\operatorname{rank}(y) = 1$ and $b_2 \leq r(X)$. Using the observation in the beginning of the proof, we get that $a_2C_1b_2 \subseteq Y$ and thus $C_1b_2 \subseteq Y$. This implies that $r(X)b_2 = 0$ which is a contradiction. Thus $QE(X) = 0$, and if we put $b = r(X)$, $a = I$ then $X = C_1b$, $Y = C_1(I-b)$ and formulas (7.11) and (7.12) hold. □

Duality gives the analogous result for C_∞:

<u>Corollary 7.8:</u> Let P be a contractive projection in C_∞ so that $Q = I-P$ is also contractive. Then P and Q act either by formulas (7.11) and (7.12), or by one of the formulas (7.13) or (7.14).

In our work we have used <u>complex scalars</u>, and our results depend on this assumption. Clearly, every contractive projection in either C_1 or C_∞ over the complex scalars is also contractive over the real scalars. The converse of this statement is not true, since as we shall presently see in the case of real scalars there exist contractive projections in C_1 and C_∞ which are not of the form described by Theorems 2.14, 2.15 and 2.16. The structure of a general contractive projection in C_1 or C_∞ over the real scalars seems to be far more complicated than over the complex scalars.

Example 7.9 Let $0 \leq \lambda, \mu \leq 1$ with $\lambda+\mu = 1$, and define

$$(7.16) \quad x = \begin{pmatrix} 1 & 0 \\ 0 & 0 \end{pmatrix} \qquad y = \begin{pmatrix} 0 & \lambda \\ \mu & 0 \end{pmatrix}.$$

For every choice of (real or complex) scalars s and t we get easily that

$$(7.17) \quad \|sx+ty\|_1 = (|s|^2+|t|^2)^{1/2} .$$

Let $X_{\lambda,\mu} = \mathrm{span}\{x,y\}$ be identified naturally as a subspace of C_1. By Theorem 2.14 it follows easily that in the case of complex scalars $X_{\lambda,\mu}$ is the range of a contractive projection from C_1 if and only if $\lambda\cdot\mu = 0$. Nevertheless, in the case of real scalars, it is easily verified that

$$(7.18) \quad P\begin{pmatrix} b_{1,1} & b_{1,2} \\ b_{2,1} & b_{2.2} \end{pmatrix} = (b_{1,1}-b_{2,2})x + (b_{1,2}+b_{2,1})y$$

defines a contractive projection from $C_1^{2,2}$ (and thus from C_1) onto $X_{\lambda,\mu}$, for every λ,μ. When performing the computations it is helpful to use the following formula for the C_1-norm of a 2×2 matrix A:

$$(7.19) \quad \|A\|_1 = (\mathrm{trace}(A^*A) + 2|\det A|)^{1/2}$$

Example 7.10 Let $0 < \alpha < 1$, and define

$$(7.20) \quad x = \begin{pmatrix} 1 & 0 \\ 0 & \alpha \end{pmatrix} \qquad y = \begin{pmatrix} 0 & 1 \\ -\alpha & 0 \end{pmatrix}$$

If A is a 2×2 matrix then

$$(7.21) \quad \|A\|_\infty = \tfrac{1}{2}(\mathrm{trace}(A^*A) + ((\mathrm{trace}(A^*A))^2 - 4|\det A|^2)^{1/2})^{1/2}$$

Thus, for every real numbers s and t we get:

$$\|sx+ty\|_\infty = (s^2+t^2)^{1/2} \tag{7.22}$$

Define

$$P\begin{pmatrix} b_{1,1} & b_{1,2} \\ b_{2,1} & b_{2,2} \end{pmatrix} = b_{1,1}x + b_{1,2}y \tag{7.23}$$

Then, it is easily verified that P is a contractive projection from $C_\infty^{2,2}$ (and thus from C_∞) onto $Y_\alpha = \mathrm{span}\{x,y\}$, provided we regard C_∞ as a space over the reals.

In the case of complex scalars however there is no contractive projection from C_∞ onto Y_α. In order to prove this it is clearly enough to prove that Y_α is not isometric to ℓ_∞^2 or to ℓ_2^2 (since every two dimensional subspace of C_1 over the complex scalars which admit a contractive projection from C_1 is isometric to ℓ_1^2 or to ℓ_2^2). We have

$$\|x+iy\|_\infty^2 + \|x-iy\|_\infty^2 = 4(1+\alpha^2) > 4 = 2(\|x\|_\infty^2 + \|iy\|_\infty^2). \tag{7.24}$$

So in Y_α the parallogram identity is false, and thus Y_α is not isometric to ℓ_2^2. If Y_α were isometric to ℓ_∞^2, we would have elements $u,v \in Y_\alpha$ with $\|u\|_\infty = \|v\|_\infty = 1$ and $u \perp v$. In particular $\det u = \det v = 0$. But, as is easily checked, the only pair $\{u,v\}$ of linearly independent elements of Y_α which satisfy $\det u = \det v = 0$ is (up to a multiplication by scalars) $u = x+iy$, $v = x-iy$. But for these u and v we have

$$u^*v = \begin{pmatrix} 1 & i\alpha \\ -i & \alpha \end{pmatrix}\begin{pmatrix} 1 & -i \\ i\alpha & \alpha \end{pmatrix} = \begin{pmatrix} 1-\alpha^2 & i(\alpha^2-1) \\ i(\alpha^2-1) & \alpha^2-1 \end{pmatrix} \neq 0 \tag{7.25}$$

contradicting $u \perp v$.

BIBLIOGRAPHY

1. J. Arazy and Y. Friedman, "The isometries of $C_p^{n,m}$ into C_p", Israel J. Math. 26 (1977), 151-165.

2. C. Bessaga and A. Pelczynski, "On extreme points in separable conjugate spaces", Israel J. Math. 4 (1966), 262-264.

3. R.G. Douglas, "Contractive projections on an L_1-spaces", Pacific J. Math. 15 (1965), 443-462.

4. A. Gleit and A. Lazar, "Basis constants for the spaces of $n\times n$ matrices", J. Funct. Anal. 22 (1976), 354-365.

5. I.G. Gohberg and M.G. Krein, "Introduction to the theory of linear non-selfadjoint operators", Amer. Math. Soc. Translations, Vol. 18.

6. R.V. Kadison, "Isometries of operator algebras", Ann. of Math. 54 (1951), 325-338.

7. Ch. A. McCarthy, "C_p", Israel J. Math. 5 (1967), 249-271.

8. R. Schatten, "Norm ideals of completely continuous operators", Springer-Verlag, Berlin, 1960.

Institute of Mathematics, The Hebrew University of Jerusalem, Jerusalem, Israel.

Department of Mathematics, Tel-Aviv University, Tel-Aviv, Israel.